AAPS Introductions in the Pharmaceutical Sciences

Volume 8

The *AAPS Introductions in the Pharmaceutical Sciences* book series is designed to support pharmaceutical scientists at the point of knowledge transition. Springer and the American Association of Pharmaceutical Scientists (AAPS) have partnered again to produce a second series that juxtaposes the *AAPS Advances in the Pharmaceutical Sciences* series. Whether shifting between positions, business models, research project objectives, or at a crossroad in professional development, scientists need to retool to meet the needs of the new scientific challenges ahead of them. These educational pivot points require the learner to develop new vocabulary in order to effectively communicate across disciplines, appreciate historical evolution within the knowledge area with the aim of appreciating the current limitations and potential for growth, learn new skills and evaluation metrics so that project planning and subsequent evolution are evidence-based, as well as to simply "dust the rust off" content learned in previous educational or employment settings, or utilized during former scientific explorations. The *Introductions* book series will meet these needs and serve as a quick and easy-to-digest resource for contemporary science.

Jenny Ka Wing Lam • Philip Chi Lip Kwok
Editors

Respiratory Delivery of Biologics, Nucleic Acids, and Vaccines

Springer

aaps

Editors
Jenny Ka Wing Lam
Department of Pharmaceutics
UCL School of Pharmacy
University College London
Brunswick Square, London, UK

Philip Chi Lip Kwok
Advanced Drug Delivery Group, Sydney Pharmacy School, Faculty of Medicine and Health
The University of Sydney
Camperdown, NSW, Australia

ISSN 2522-834X ISSN 2522-8358 (electronic)
AAPS Introductions in the Pharmaceutical Sciences
ISBN 978-3-031-47566-5 ISBN 978-3-031-47567-2 (eBook)
https://doi.org/10.1007/978-3-031-47567-2

Editorial Contact: Charlotte Nunes

This Springer imprint is published by the registered company Springer Nature Switzerland AG
The registered company address is: Gewerbestrasse 11, 6330 Cham, Switzerland

Paper in this product is recyclable.

Preface

Biologics have emerged as an increasingly important class of therapeutics. Owing to their high specificity and ability to target specific components of the immune system, biologics are particularly useful in the treatment of cancers and immune-mediated diseases. In 2022, biologics accounted for over 40% of approvals by the Centre for Drug Evaluation and Research (CDER) of the United States Food and Drug Administration, a record high percentage of biologics to date. Although not classified as biologics, nucleic acids such as small interfering ribonucleic acid (siRNA) and messenger ribonucleic acid (mRNA) also gained much attention in recent years. Their capability to manipulate the expression of virtually any gene makes them remarkably powerful therapeutic modalities to exploit "undruggable targets" to treat a wide range of diseases. The COVID-19 pandemic that began in 2020 has accelerated the development of new vaccine platforms that prominently feature viral vectors and mRNA vaccines. The landscape of pharmaceutical development is rapidly evolving, so biologics are expected to overtake small molecules and dominate the market in the foreseeable future.

Despite many biologics and vaccines already approved for treating or preventing lung diseases such as lung cancer, asthma, and respiratory viral infections, nearly all of them are parenterally administered. In contrast, administration through inhalation can directly target therapeutics to the site of action in the airways to maximise efficacy. This non-invasive route overcomes many drawbacks associated with parenteral routes, such as the requirement of trained personnel for administration, sterile formulations, and the risks of needlestick injuries and blood-borne diseases. Intranasal and orally inhaled vaccines can trigger mucosal immune response, which is lacking in conventional injected vaccines, and provide better protection against respiratory infections. Moreover, the high vascularisation of the lungs and nasal cavity can serve as an entry portal for the systemic delivery of small proteins and peptides. The potential of biologics, nucleic acids, and vaccines could be expanded to benefit more patients by developing formulations beyond parenteral administration, as the orally inhaled and nasal routes are already well-established for small drug molecules against respiratory diseases.

This book consists of nine chapters which give a comprehensive overview of the challenges and strategies in delivering a wide range of biologics and macromolecules to the respiratory tract, including antibodies, proteins, peptides, nucleic acids, and vaccines. It also covers the respiratory delivery of probiotics and bacteriophages, which have not yet been extensively studied but have shown promising potentials in improving health and fighting against infections. The pharmacokinetics of inhaled biological macromolecules, which is a poorly understood topic, are also discussed. We hope this book provides invaluable insights for readers who are working towards the goal of developing orally inhaled and nasal biologics, nucleic acids, and vaccines.

Brunswick Square, London, UK — Jenny Ka Wing Lam
Camperdown, NSW, Australia — Philip Chi Lip Kwok

Contents

About the Editors

Jenny Ka Wing Lam is an Associate Professor of Pharmaceutics at the UCL School of Pharmacy, University College London. Dr Lam obtained her MPharm and PhD from The University of Nottingham, UK. In her PhD study, she investigated the use of non-viral vectors for gene delivery. Dr Lam was then awarded the Maplethorpe Fellowship and continued her research in nucleic acid delivery in the Department of Pharmacy at King's College London. She then joined the Department of Pharmacology and Pharmacy in the University of Hong Kong as Assistant Professor, and returned to the UK in 2022. Dr Lam's research is focused on the development of novel delivery system for a wide range of therapeutics including small molecules, nucleic acids, and biologics, with special interest in the use of particle engineering methods to produce aerosol formulations for the treatment/prevention of respiratory diseases. She is the Principal Investigator of a number of competitive grants. She has published over 90 peer-reviewed articles and filed a number of patent applications on pulmonary drug delivery systems. In 2020, she was awarded the DDL (Drug Delivery to the Lung) Emerging Scientist Award which recognised her significant accomplishment and innovation in inhalation science.

Philip Chi Lip Kwok is a Senior Lecturer in Pharmaceutical Sciences at the School of Pharmacy, The University of Sydney. He obtained his Bachelor of Pharmacy degree with First Class Honours in 2002 from this Faculty. He became a registered pharmacist after one year of training in a community pharmacy. Dr Kwok then undertook his PhD studies on pharmaceutical aerosol electrostatics in the Faculty of Pharmacy at The University of Sydney and graduated in 2007. He was a Research Associate in the same group until August 2011 and became an Assistant Professor in the Department of Pharmacology and Pharmacy at The University of Hong Kong in September 2011. Dr Kwok returned to The University of Sydney as a Lecturer in Pharmaceutical Sciences at the end of July 2017. His research is in pulmonary drug delivery. In particular, he specialises in the engineering, physico-chemical characterisation, and electrostatics of pharmaceutical aerosol formulations. He has collaborated with academic and industrial researchers, both locally and internationally, on formulation-focused as well as interdisciplinary projects.

Design Strategies of Dry Powders for Pulmonary Delivery of Pharmaceutical Peptides

Hideyuki Sato

Abstract Pharmaceutical peptides have gradually become more attractive therapeutic molecules than small-molecule drugs, since pharmaceutical peptides are more selective and effective, and have fewer side effects, compared to small-molecule drugs. Generally, owing to their poor oral absorbability and gastrointestinal stability, peptides are mainly administered via intravenous and intramuscular routes, which adversely affects patient compliance. Pulmonary characteristics, such as large surface area, abundant capillary network, thin membrane with adequate permeability for macromolecules, reduced enzymatic degradation, and lack of first-pass metabolism, facilitate the use of inhalable formulations to achieve local and systemic actions. Precise control of powder properties and appropriate design of respirable particles to stabilize target peptides are necessary for their efficient pulmonary delivery. This chapter discusses the strategies for formulation and efficient delivery of peptide-loaded dry powders.

Keywords Dry powders · Particle design · Peptides · Pulmonary delivery

1 Introduction

Recent advances in formulation development, such as drug engineering and recombinant DNA technology, facilitate the efficient formulation and delivery of medium and high molecular weight compounds such as peptides, nucleic acids, antibodies, and proteins, which have specific and potent pharmacological actions owing to their high target selectivity. In 2019, eight biologic drugs were among the top 10 global sales [10]. Recently, middle molecular weight drugs with a molecular weight of

H. Sato (✉)
Laboratory of Biopharmacy, School of Pharmaceutical Sciences, University of Shizuoka, Shizuoka, Japan
e-mail: h.sato@u-shizuoka-ken.ac.jp

J. K. W. Lam, P. C. L. Kwok (eds.), *Respiratory Delivery of Biologics, Nucleic Acids, and Vaccines*, AAPS Introductions in the Pharmaceutical Sciences 8,
https://doi.org/10.1007/978-3-031-47567-2_1

about 500–6000 Da, peptides and nucleic acids, have attracted significant interest as they are relatively easy to design, synthesize, and control, compared with proteins and antibodies. However, similar to other biologics, the oral bioavailability of peptides is poor, owing to their molecular size, degradation by digestive enzymes in the gastrointestinal (GI) tract, low permeability through the epithelial barriers in the GI tract, and first-pass metabolism. Thus, most peptide drugs are administered via the parenteral routes, like subcutaneous, intramuscular, or intravenous injections. Parenteral administration has several limitations, such as pain, risk of injury at the injection site, difficulties with self-administration, cold chain storage, production of needles, syringes, and other waste materials that are difficult to dispose of. To overcome these drawbacks, safe, effective, and non-invasive routes and drug delivery technologies are being developed.

Pulmonary delivery can offer rapid absorption of drugs and higher systemic exposure because of the extensive vascularization in the lung, high tissue permeability of epithelial cells, relatively low activity of metabolic enzymes, and large pulmonary surface area [57], leading to several advantages over conventional, non-invasive administration routes in the treatment of systemic diseases. The pulmonary administration system can deliver a larger amount of drug directly to a local disease site with minimal systemic exposure. There have been many reports on achieving efficacious drug delivery to the disease site for the treatment of asthma, chronic obstructive pulmonary disease, and respiratory infections [28]. Therefore, pulmonary delivery is a viable option for effective and safe delivery of therapeutic peptides for topical and systemic actions, depending on the target disease. Thus, much attention has been paid to develop and investigate pulmonary delivery systems for therapeutic peptides. For the systemic actions, pulmonary delivery of metabolic hormones, including insulin, calcitonin, growth hormones, somatostatin, thyroid-stimulating hormone, and follicle-stimulating hormone, to humans and experimental animals has been reported [1]. For the treatment of respiratory diseases like asthma, chronic obstructive pulmonary diseases, and cystic fibrosis and lung infections, applications of vasoactive intestinal peptide analogues, neuropeptide Y, cyclosporine A, and colistin have been investigated [7, 16, 79]. However, difficulty in particle design and formulation strategies limits the practical applications of pulmonary delivery of therapeutic peptides. Generally, for orally inhaled particles, aerodynamic particle size between 1 and 5 μm ensures efficient delivery of inhaled particles [13]. Particles with aerodynamic diameters larger than 10 μm are deposited in the extrathoracic sites, such as the mouth, pharynx, and larynx, and particle sizes between 5–10 μm are deposited in the tracheobronchial tree. Particles that are too small cannot adhere to the surface of the respiratory tract because they are removed from the respiratory tract by exhaled airflow. Compared with conventional small-molecule drugs, therapeutic peptides tend to be at risk of degradation and deactivation, owing to their poor stability against chemical and physical stress during the production process and storage [28]. In addition, even after reaching the respiratory site, some physiological factors cause their degradation/denaturalization and elimination from the target site [30]. Thus, along with the precise control of particle size, the selection of appropriate excipients and formulation design are key considerations for the successful pulmonary delivery of therapeutic peptides.

There are three commonly used systems for inhalation medication: nebulizers, metered dose inhalers, and dry powder inhalers (DPIs). Among these, DPIs are the most commonly used inhalation devices in adults patients due to a lot of practical advantages including ease of use, portability, and relatively high pulmonary delivery efficiency [70]. Additionally, considering the chemical stability of peptides, powderization would be one of the options to stabilize the target molecule, since the presence of water could accelerate the degradation of peptides. Thus, this chapter mainly focuses on the physicochemical and biological challenges of therapeutic peptides and particle design/formulation strategies for DPIs to achieve efficient peptide delivery by oral inhalation.

2 Factors Affecting Peptide Stability in Manufacturing Process

During the manufacturing process of inhalable formulations, many factors cause the degradation, aggregation, and/or deactivation of target peptides due to preparation conditions and chemical and physical stresses such as thermal stress, oxidative stress, shear stress, and pressure. Deamidation, covalent aggregation, oxidation, and Maillard reaction are the main degradation pathways [77]. Additionally, in some cases, aggregation is irreversible and reduces the physical stability of the peptide, not only leading to a loss of activity but also other critical problems such as toxicity and immunogenicity [54]. Thus, to prevent the risk of loss of quality and toxicity of the formulation, appropriate conditions and excipients should be selected. This section briefly summarizes general information on the potential factors influencing peptide stability (Table 1).

Table 1 Factors affecting peptide stability in manufacturing process

Factor	Comment
pH	Solution pH can influence the charge state of dissolved peptide depending on the isoelectric point (pI), changing the dispersion state by electric repulsion.
Physical stress (heat, pressure, and shear stress)	Physical stresses in manufacturing process can alter molecular interactions including electrostatic force, hydrophobic interactions, hydrogen bonding, van der Waals forces, and local peptide interactions, possibly changing the folding state and causing aggregations.
Oxidative stress	Amino acids with reactive side chain in a peptide (histidine, methionine, cysteine, tyrosine, and tryptophan) can be potentially damaged by reaction with any of a number of reactive oxygen species during formulation process and storage.
Excipient	Buffering agents and salts have potential to influence on peptide stability due to changing pH and ionic strength of peptide solution. Some excipients like surfactants and antioxidants can stabilize peptide.

2.1 pH

Generally, the pH can influence the aggregation and chemical degradation of peptides. Electrostatic interactions are known to play a major role in stabilizing the dispersion state. More specifically, a higher net charge can contribute to the suppression of the aggregation potential of peptides owing to the electronic repulsion between peptide molecules [60]. Additionally, the ionic strength and nature of the cations and anions in solution, and the presence of polyelectrolytes also have the potential to affect the rate and extent of aggregation. To estimate the effect of solution conditions on aggregation, the isoelectric point of peptides, where net charges are essentially zero, is useful information [38], suggesting the importance of selecting suitable excipients for stabilizing target peptides in the formulation process and subsequent products. When the solution pH is significantly different from the isoelectric point, the protein becomes highly charged, resulting in electric repulsions.

2.2 Temperature

During the pharmaceutical process, active ingredients experience physical stress, such as thermal stress, pressure, and shear stress. Temperature affects the structural stability of peptides via molecular interactions, conformational stability (secondary structure), solubility, and chemical degradation. Electrostatic forces, hydrophobic interactions, hydrogen bonding, van der Waals forces, and local peptide interactions are responsible for the free energy of the folding state. Temperature plays a crucial role in reaction kinetics because rate constants increase exponentially with temperature, i.e., the rate of aggregation increases at high temperatures because of the rise in global molecular mobility (kinetic energy) [6]. Therefore, an increase in temperature accompanies an increase in the probability and number of collisions with sufficient energy to overcome the activation energies for the reaction, possibly leading to the acceleration of aggregation formation. Consequently, they may compromise their therapeutic efficacy and cause potential safety concerns. The lyophilization/freeze-drying technique is a very common method to increase both chemical and physical stability [18] and reduce the thermal stress for solidification. However, in some cases, these processes result in conformational changes in peptides, leading to increased aggregation after reconstitution in water, depending on their physicochemical characteristics. Considering these points, the thermal stress during drying and evaporation conditions should be carefully selected to minimize the risk of aggregation, and even using freeze-drying, the possible changes in conformational changes and redispersibility of the peptide should be evaluated by spectroscopic analysis such as circular dichroism spectroscopy.

2.3 Oxidative Stress

Oxidation of peptides is one of the major problems associated with chemical and physical instability. It is well known that some amino acid chains can be oxidatively modified during peptide purification, formulation, and storage [47, 71]. These peptides included aromatic amino acids (Tyr, Trp, and His) and sulphur-containing side chains (Met and Cys). Thus, there have been many reports on the application of antioxidants, such as ascorbic acid, for stabilization [35]. Methionine, catalase, sodium thiosulfate, and chelating agents such as ethylenediaminetetraacetic acid (EDTA) and diethylene triamine pentaacetic acid (DTPA) have also been used as stabilizers against oxidative stress [40, 80]. Replacement of these sensitive amino acids with other chemical moieties may increase the chemical stability of peptide. In this case, the conformational changes and affinity to the target site for pharmacological action should be carefully evaluated after modification of the amino acid sequence and/or chemical moieties of target peptide.

2.4 Excipients

Different excipients are used in pharmaceutical development of peptide formulations, including salts, polyols, sugars, surfactants, osmolytes, chelators, antioxidants, specific ligands, and carbohydrates [77]. As mentioned above, pH conditions can strongly affect the stability and potential of aggregation. Therefore, buffering agents such as acetate, citrate, histidine, phosphate, and Tris can alter stability. Salts also have a complicated influence on peptide stability by altering both conformational and colloidal stability [35]. Depending on the difference in surface charge, the overall effect of a salt on physical stability is a balance of different and multiple mechanisms by which the salt interacts with water and biomolecules (Hofmeister effects and Debye–Hückel effects). Using a surfactant, the exposure of the hydrophobic part of the peptide to water is minimized owing to its amphiphilic properties, resulting in the stabilization of the peptide. Pharmaceutical excipients have both positive and negative effects on the stability of peptides, suggesting the necessity of careful selection depending on the physicochemical properties of the target peptides.

3 Biological Barriers

Active ingredients can act in the mucus layer or on pathogens for topical pharmacological actions in the respiratory tract, whereas for systemic delivery, target drugs have to permeate the bronchial mucus layer and alveolar epithelial cells to enter systemic circulation (Fig. 1). The thickness of the mucus layer is between 5 and 55 μm depending on the depth of the site in the respiratory tract [29] and varies

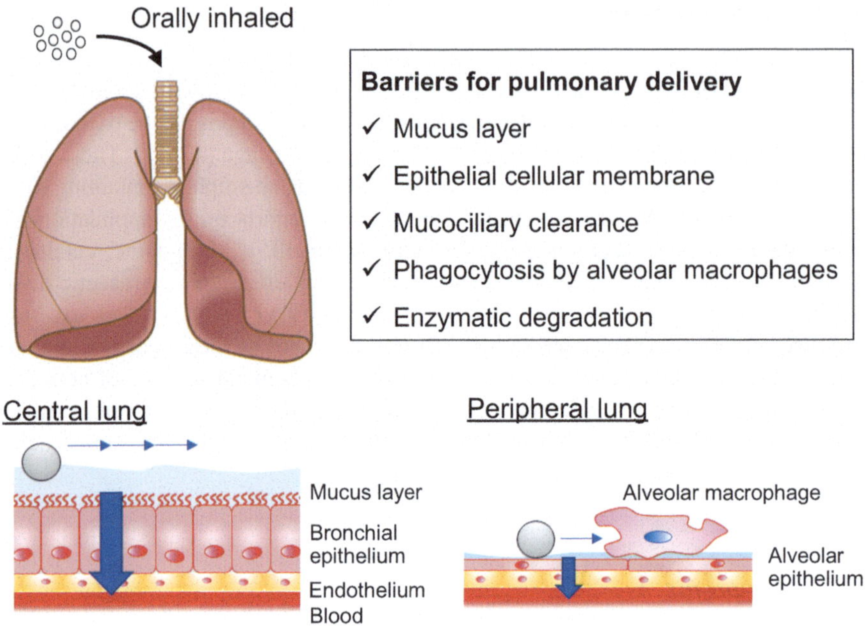

Fig. 1 Biological barriers for pulmonary delivery of peptide drugs

depending on the physiological conditions [27], resulting in the difficulty of sufficient delivery of target peptides in obstructive diseases, which causes increases in mucus production and viscosity of mucus. Thus, controlling diffusiveness within the mucus layer by mucoadhesive and mucopenetrating properties of the formulation could contribute to the enhanced and/or sustained absorbability from the lung (described in detail in Sect. 5.1 Adjusting mucodiffusiveness). Although pulmonary mucus is not a significant barrier for macromolecules with sizes less than 10 nm (−500 kDa), aggregates of peptides and nano/microparticle formulations (>10 nm) could be trapped by the mucus layer owing to the mesh-like structure of the mucus layer. They are transported from the deep lung to the pharynx by mucociliary clearance and are removed via phagocytosis by macrophages.

For systemic action of the target peptides, they must penetrate the epithelial layer. For permeation via passive diffusion, there is a suitable range of lipophilicity (log *P* value: 2–9) and a cutoff value for molecular size (above 1000 Da) [4, 48]. Thus, they can permeate the cellular membrane via transcytosis (which may be receptor-mediated or carrier-mediated), paracellular routes, and large transitory pores in the epithelial layer. Proton-coupled peptide transporters (PEPT) are generally known as the main transporter for the permeation of di-and tripeptides; however, the contribution of PEPT to the absorption of inhaled peptides in the pulmonary tract is not significant. If target peptides can be captured by endocytosis and pinocytosis during the absorption process, they can be found in lysosomes, possibly leading to enzymatic degradation by hydrolases in the lysosomes. Therefore, appropriate strategies are required to avoid inactivation by enzymatic degradation.

In the respiratory tract, the activities of metabolic enzymes are lower than those in the gastrointestinal tract and liver tissues, and molecules absorbed via pulmonary delivery can avoid first-pass metabolism, suggesting great advantages for peptide delivery via inhalation systems. However, degradation of peptides and proteins also occurs in the respiratory tract, and the extent of degradation depends on the molecular size. Relatively small peptides with molecular sizes of less than 3 kDa are more sensitive than larger proteins (6–500 kDa) [73]. In bronchoalveolar lavage fluid, angiotensin-converting enzyme, cathepsin D, cathepsin H, and dipeptidyl peptidase IV are abundant enzymes involved in the degradation of peptides and proteins.

Proteins and peptides deposited on ciliated epithelium are not significantly absorbed owing to mucociliary transportation up the airways and movement into the gastrointestinal tract, resulting in the degradation and denaturation of the molecules. The clearance mechanism in the alveolar region includes phagocytosis by macrophages, paracellular diffusion through tight junctions, vesicular endocytosis or pinocytosis, and receptor-mediated transcytosis. For soluble proteins, clearance via phagocytosis by alveolar macrophages does not seem important as a clearance system because macrophages preferentially capture relatively insoluble particles.

Some respiratory diseases, including asthma, chronic obstructive pulmonary diseases, lung fibrosis, and pulmonary hypertension, cause physiological changes such as bronchial constriction, emphysema, and increase of mucus thickness and viscosity. It has also been reported that the activities of several proteases are altered due to the migration and accumulation of inflammatory cells at the inflammatory site and the release of protease from inflammatory cells [22]. Thus, peptide degradation increases in most lung diseases. An increase in the thickness of the mucus layer in the diseased state is also a non-enzymatic barrier for the delivery of peptides to target cells. In patients with cystic fibrosis, a combination of increased mucus production and viscosity can be observed, indicating the difficulty of pulmonary drug delivery compared with the healthy state [27]. Additionally, tissue remodelling caused by fibrosis in COPD and idiopathic pulmonary fibrosis evokes an increase in the air-blood barrier thickness due to tissue hypertrophy.

4 Production and Design of Peptide-Load Inhalable Particles for DPI

To manufacture inhalable particles with desirable inhalation performance and sufficient peptide delivery, the production method and particle design are crucial for determining the potential of the DPI system. The particle manufacturing approach can be categorized using an intuitive approach based on whether the starting

Table 2 Particle production methods for inhalable powders

Category	Technique	Characteristic	Reference
Top-down approaches			
Milling	Ball/beads milling	High shear stress Necessity of separating solid milling media Concerns of contamination with foreign particulate	
	Jet milling	Commercially established method for small molecular drugs for inhalation No solid milling media No subsequent separation process	[33, 53, 78]
Homogenization	High-pressure homogenization	Necessity of evaporation step after micronization Commercially used for non-inhalation drugs	
Bottom-up approaches			
Solvent evaporation	Spray-drying	Commercially established method as single-step particle formation process Controlling size, morphology, density, surface composition	[3, 8]
	Spray freeze-drying	Variant of spray-drying process Two-step process (freezing and lyophilization) Production of fragile particles with very low density No thermal stress, but poor scalability and complex process	[58, 59, 76]
	Supercritical fluid method (rapid expansion): Supercritical fluid-assisted atomization	Use SCF CO_2 as atomizing medium No use of organic solvent	[31, 64]
Antisolvent	Supercritical fluid method (antisolvent)	Use SCF CO_2 as an anti-solvent to precipitate fine particles	[36, 37]

material is a solid particle (top-down method) or a liquid (bottom-up method) (Table 2). The manufacturing approach and composition have a significant impact on the powder properties and the physical and chemical stability of the included peptides. Thus, powderization techniques should be carefully selected based on the physicochemical properties and stability of the target peptide.

4.1 Preparation Method

4.1.1 Top-Down Method

In top-down approaches, target bulk particles are mechanically broken down into micron-sized particles or aggregated in the respirable size range (e.g., milling and high-pressure homogenization). Among the top-down approaches, jet milling is a well-established method for the production of DPI systems because it is faster, more scalable, and offers a better-controlled particle size distribution than homogenization or ball milling. Additionally, because of the media-less milling process, there is no requirement for the separation of micronized particles from the milling media. During jet milling, coarse particles of the raw feed are injected into the milling chamber with a forceful stream of high-pressure gas and then pulverized by repetitive inter-particle collisions and attritions [41]. However, similar to other top-down approaches, there are stability concerns due to the susceptibility of macromolecules in the environment, and the heat and mechanical stresses generated by high-energy conditions and harsh physical grinding. High pressure is also known to cause denaturation of peptides and proteins [77]. Depending on the operating conditions of jet milling, the milling process contributes to the loss of potency of target peptides due to the mechanical stresses of the jet-milling process. Although top-down approaches are commonly used for the micronization of many types of small-molecule drugs, optimized micronizing conditions should be considered in the case of biologics such as peptides.

4.1.2 Bottom-Up Method

In contrast to top-down approaches, there are various bottom-up approaches, leading to the diversity of starting materials, including simple solutions, cosolvent solutions, oil in water (O/W) or water in oil (W/O) emulsions, suspensions, or more complex colloidal fluids, suggesting flexibility for designing functional particles. The particles are generated by single-step solvent evaporation, such as spray-drying, and more complex processes, such as precipitation using an antisolvent. Compared to top-down approaches, the molecular, colloidal, or powder nature of bottom-up approaches can be well-controlled, such as particle morphology and size, surface properties, and crystallinity. Additionally, the combined use of pharmaceutical excipients in bottom-up approaches allows further characterization of the physicochemical, chemical, and physical stability and inhalation performance of DPI.

Spray drying is a well-established powderization technology for the preparation of solid particle formulations [24]. This process consists of three steps: atomization of the sample solution into fine droplets, drying by hot air, and collection of the generated powder. This technique has been used to develop formulations of small-molecule drugs with poor water solubility; however, it is increasingly used to provide powder formulations of macromolecules and biopharmaceuticals [24]. The freeze-drying method is commonly used for solidification of peptide solutions. Compared with the freeze-drying method, the spray-drying method can generate fine particles with fine aerosolization performance for DPI and can control the powder properties in terms of morphology, size distribution, shape, and surface texture with scalability. During the spray-drying process, it is possible to expose some stresses, including mechanical stress, heat stress, and adsorption at the air-liquid interface during atomization. Although spray-drying has fewer sources of stress to induce denaturation of peptides than the mechanical top-down approach, there is still some possibility of aggregation and denaturation of peptides during the process. Thus, the atomizing conditions (atomizing pressure, feed pump, and feed concentration) and drying conditions (temperature of hot air and flow rate) should be carefully optimized depending on the target peptides.

Spray-freeze-drying is a variant of the spray-drying process used to generate highly porous particles, such as freeze-dried cakes. In this process, the droplets generated from the spray nozzles were directly frozen using liquid nitrogen and then lyophilized [45]. This process can be used to prepare inhalable spherical particles with high porosity, thereby improving inhalation performance. However, the spray-freeze-drying process is a very complex process with poor scalability; thus, the applicability might be limited in some cases using highly heat-sensitive compounds in conventional spray-drying systems.

Supercritical fluid (SCF) technologies have also been applied to prepare particles for inhalation [75]. Owing to relatively mild conditions (relatively low critical temperature: 31.1 °C and pressure: 7.38 MPa), SCF carbon dioxide (SC-CO_2) is commonly used to develop the formulations. During the particle production process, SC-CO_2 can act as an anti-solvent to precipitate peptide-loaded particles. This process has been used to prepare dried protein and peptide formulations, and to provide the possibility of producing small microparticles suitable for inhalation. Briefly, an aqueous sample solution containing peptides, proteins, and ethanol was atomized through a coaxial nozzle into a pressurized chamber filled with SC-CO_2. The drug-loaded particles were then precipitated from the water droplet. Among the various SCF-based powderization systems, SCF-assisted atomization (SAA), where SC-CO_2 serves as the atomizing medium, has been highlighted for its availability in aqueous solutions without the use of any organic solvents [64]. Although there are still no products for commercially available DPI manufactured by SCF systems, some studies on the production of macromolecule-loaded powders, including insulin, lysozyme, and albumin, have been reported [17, 23, 61, 64]. According to these studies, although biologics-loaded particles can be successfully developed by the SCF approach, the stability and inhalation performance of the prepared particles were dependent on the physicochemical characteristics of the target peptides and process conditions, such as temperature and pH conditions.

In addition to the conventional technique to prepare uniform micron-sized particles for inhalation, some unique technologies have been reported recently. The fine droplet drying (FDD) process is a powderization technique that employs an inkjet head used in the printing industry [62, 68]. In this process, the inkjet head uses a piezo element as an actuator to produce uniform fine droplets, resulting in uniform particles after drying the generated droplets. The size of the produced particles could be precisely controlled at the single-micron scale by changing the size of the nozzle holes. In a previous study, salmon calcitonin was encapsulated in poly(lactic-*co*-glycolic acid) using the FDD process for inhalation to sustain its pharmacological action [62].

4.2 Excipients to Enhance Stability of Peptides

As mentioned above, there are many possible stresses that induce the degradation, aggregation, and denaturation of target peptides during the manufacturing process of inhalable particles. The target compounds should be protected from such stresses and stabilized without affecting their biological activity. Immobilization of peptides by solidification and entrapment within the carrier matrix can simply contribute to preserving biological action to prevent aggregation and conformational changes. There are many kinds of excipients, such as sugars, polyols, salts, amino acids, polymers, and surfactants, which can be applied as stabilizing agents for peptides during the manufacturing process (Table 3). However, the use of excipients for stabilization of peptide-loaded DPI systems is more difficult than that of oral formulations because only a limited range of compounds is approved as a pharmaceutical excipient for pulmonary administration.

Sugars and polyols are typically used as pharmaceutical excipients to improve the flowability and stability of peptides. Sucrose, mannitol, lactose, trehalose, sorbitol, and inulin are common excipients used for the production of peptide-loaded particles [21]. These sugars and polyols can theoretically stabilize the included peptides based on two theories: (i) water replacement theory and (ii) vitrification theory (Fig. 2). Water replacement theory suggests that during the solidification/powderization process, these excipients can form hydrogen bonds with active ingredients by replacing the hydrated water of protein/peptide, leading to the entrapment of macromolecules into the matrix structure of excipients. This may reduce the chance of hydrolysis by removing surrounding water. Vitrification theory suggests that entrapment of target macromolecules into the glass-forming matrix former can contribute to the restriction of mobility in a rigid structure, resulting in the stabilization of highly structured proteins and peptides. Glassy excipients with a high glass transition temperature can act as physical barriers to improve thermostability. Non-reducing sugars, including trehalose and sucrose, are preferable, and the combined use of sugars, polyols, and other stabilizing excipients can stabilize proteins and peptides during the manufacturing process.

Amino acids are highly biocompatible and this class of compounds have a wide range of chemical and physicochemical properties, including hydrophilic, hydrophobic, neutral, cationic, anionic, and antioxidant effects [32]. Amino acids with relatively small sizes, including glycine, alanine, leucine, isoleucine, histidine, and arginine, are typically used as stabilizers in DPI formulations. They can form hydrogen bonds with proteins and peptides, possibly leading to stabilization during

Table 3 Excipients for using stabilization of peptide and proteins during spray-drying process

Category	Excipient	Characteristic	Reference
Sugars	Glucose, fructose, maltose, sucrose, trehalose, inulin, dextran	Water replacement theory contributes to the stabilization of peptide by molecular interactions. Entrapping peptide into the glassy matrix structure of excipients, stabilizing included molecules by restricting mobility.	[14, 19, 21, 66]
Polyols	Glycerol, mannitol, sorbitol	Water replacement theory contributes to the stabilization of peptide by molecular interactions. Entrapping peptide into the glassy matrix structure of excipients, stabilizing included molecules by restricting mobility.	
Amino acids	Alanine, arginine, aspartic acid, glycine, histidine, leucine, isoleucine, lysine, phenylalanine, proline	Forming hydrogen bonding with proteins and peptides. Inhibition of the aggregation of peptides by competing with adsorption at the air–liquid interface. Improved inhalation performance and moisture protection by hydrophobic amino acids.	[2, 15, 25, 26, 32]
Surfactants	Polysorbate 20, polysorbate 80, dipalmitoylphosphatidylcholine, sodium glycocholate	Inhibition of the aggregation of peptides and assisting in the refolding by preventing adsorption at the air–liquid interface.	[11, 43, 44]

powderization. Additionally, they can inhibit the aggregation of peptides by competing with adsorption at the air-liquid interface. The application of hydrophobic amino acids, L-leucine and phenylalanine, can also reduce moisture-induced degradation and improve the inhalation performance of dry powders by acting as dispersibility enhancers. L-leucine can enhance the aerosolization efficiency of spray-dried powders by reducing their surface cohesiveness. Surfactants such as polysorbate 20, polysorbate 80, and dipalmitoyl phosphatidyl choline can also be used as stabilizers to inhibit aggregation and assist in the refolding of proteins and peptides by preventing adsorption at the air-liquid interface or ice-liquid interface during the drying process.

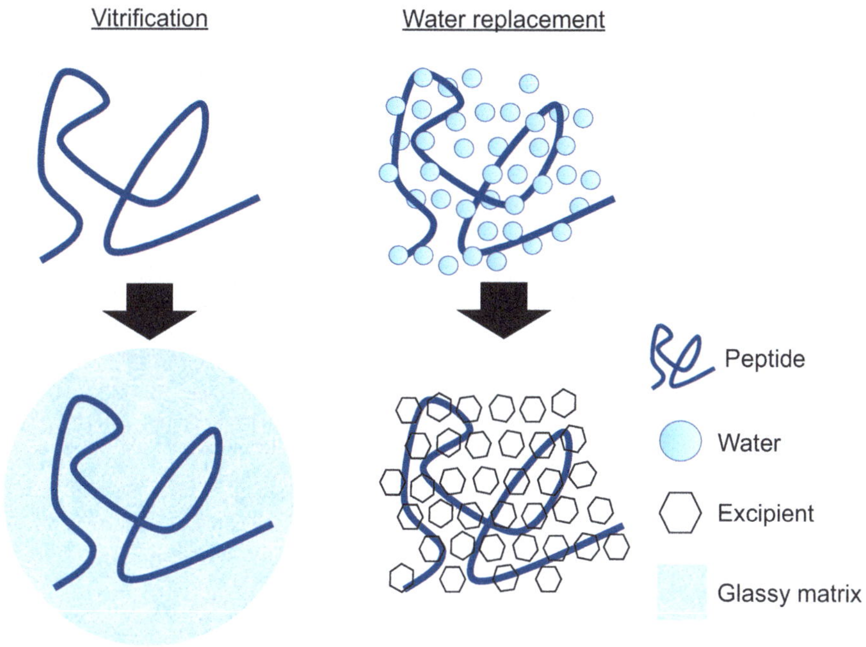

Fig. 2 Vitrification and water replacement theories of peptide stabilization by sugars and polyols

5 Pulmonary Delivery Strategies for Therapeutic Peptides

To achieve efficient delivery and an optimal therapeutic index of therapeutic peptides (also peptide-loaded carriers), pharmaceutical strategies should be carefully selected to overcome the biological barriers mentioned above. Some approaches include controlling mucodiffusiveness in the mucus layer, stabilization by molecular modification, and particle engineering of encapsulation [24, 55] (Table 4).

5.1 Adjusting Mucodiffusiveness

The pulmonary mucus layer is known as the physiological barrier that prevents the undesirable absorption and elimination of foreign substances. Mucins, the primary non-aqueous component of mucus, are polymers that have a complex and heterogeneous structure with domains that undergo various molecular interactions, such as hydrophilic/hydrophobic, hydrogen bonds, and electrostatic interactions. The permeability of the mucus layer can be mainly influenced by the size and charge of the drugs owing to the structural characteristics and components of the mucus layer [55]. In the case of a relatively small size under 10 nm (or molecularly dispersed state), the capturing efficiency of the mucus layer does not significantly influence

Table 4 Pulmonary delivery strategies of peptide drugs

Strategy	Function	Example	Reference
Controlling mucodiffusiveness	Mucopenetrating	PEGylated carrier	[9, 50, 69]
	Mucoadhesive	Mucoadhesive polymer (chitosan, alginate, hyaluronic acid, and cellulose derivatives)	[5, 46, 51]
Chemical modification/ derivatization	Stabilization	Replacement of amino acid Chemical modification Cyclization	[52, 53, 74, 56]
Encapsulation into carrier particles	Stabilization and designing functional formulations (controlled release, mucoadhesive, and mucopenetrating)	Natural polymer (gelatin, hyaluronic acid, albumin, chitosan, and carrageenan) Synthetic polymer (cellulose derivatives, acrylic acid derivatives, PLA, and PVA) Synthetic block copolymer (PLGA, PEG-PLGA, and PEG-PLA)	[20, 34, 39, 63, 65, 67]

PEG polyethylene glycol, *PEG* polylactic acid, *PEG* polyvinyl alcohol, *PLGAPEG* poly(lactic-*co*-glycolic) acids

the penetration of molecules [12] because the size of the mesh structure ranges from tens to several hundreds of nanometres. Thus, the reduction of molecular size is a possible approach to improve the delivery efficiency of large proteins. In contrast, in the case of large molecules, aggregates, and micro/nano particle formulation, controlling the interaction between mucus and modification of surface charge and PEGylation would be considered a preferable option [50]. Generally, physical entanglement and electrostatic interactions with mucin are known as the main driving forces that increase adhesiveness within mucus. The high flexibility and high hydration potential can attenuate the molecular interactions between mucus and PEG, resulting in improved diffusiveness in the mucus layer (mucopenetrating property) [69]. PEGylation also contributes to the improvement of dispersibility and protection from enzymatic degradation via steric hindrance derived from the PEG chains. Despite the advantage of PEGylation of molecules, the existence of a PEG chain would influence its pharmacological action because of steric hindrance against the pharmacophore, suggesting the necessity of considering an appropriate balance between stabilization and pharmacological actions [9]. In addition to the PEGylation of drug molecules, PEGylated carriers, especially PEG-coated nanoparticle like liposomes, and polymeric nanoparticles are an available strategy to overcome the barrier of the mucus layer [69]. It has been reported that a PEG chain with a molecular weight range of 1000–2000 Da is a suitable characteristic for drug delivery [50]. The reduction in phagocytosis by alveolar macrophages was also reported to enhance the delivery efficiency of PEG-coated macromolecule-loaded nanoparticles. Some mucoadhesive excipients have also been reported, including chitosan, alginate, hyaluronic acid, and hydroxypropyl cellulose, which enable prolonged retention of drug-loaded particles by slowing their clearance from the lungs [5, 46, 51].

5.2 *Stabilization by Chemical Modification and Derivatization*

Modification of the chemical structure of the target active ingredients is also a very important approach to improve stability under physiological conditions of the lung, possibly leading to prolonged retention and improved bioavailability of the target for not only small molecules but also macromolecules. Enzymatic degradation by peptidases in the lungs can accelerate their clearance and limit their bioavailability [28]. To avoid digestion by these enzymes, the replacement of some amino acids, modification of the amino and/or carboxy terminals, and cyclization of the molecules have been applied in previous reports [49, 74]. As described in the previous section, conjugation of PEG chain is also a common strategy for increasing stability against enzymatic degradation [56]. These modifications tend to alter the pharmacological actions owing to the conformational changes of macromolecules; thus, careful evaluation to select the site for modification is necessary (especially against the active binding site of macromolecules).

5.3 *Encapsulation into Micro/Nanocarrier Particles*

Particle engineering strategies rely on controlling or avoiding physiological clearance systems against particles, such as mucociliary clearance, cellular uptake by alveolar macrophages, enzymatic degradation, and absorption from the lung to systemic circulation. Encapsulation into micro- and nano-sized carrier particles can offer pioneering concepts for the development of optimized therapeutic tools to achieve protection from pulmonary clearance systems and controlled release of inner drugs, contributing to improved stability and prolonged topical/systemic exposure of drugs [72]. Polymeric particles and lipid-based particles, including liposomes, lipid nano/microspheres, and solid lipid nanoparticles, have been widely investigated as potential carriers for inhalable formulations [42, 65]. For the production of polymeric particles, (i) natural polymers (gelatine, hyaluronic acid, albumin, chitosan, carrageenan, etc.), (ii) synthetic polymers [cellulose derivatives, acrylic acid derivatives, poly(lactic acid)(PLA), poly(vinyl alcohol), etc.], and (iii) synthetic block copolymers [poly(lactic-co-glycolic acid)(PLGA), PEG-PLGA, and

PEG-PLA, etc.] have been strategically applied to design particles for inhalation [39, 65, 67]. In polymeric particle systems, sustained release can be achieved by controlling the diffusion of drug molecules through a polymeric matrix. Although a number of studies have evaluated polymeric particles for inhalation, there are still no approved DPI systems that contain polymeric excipients, despite their safety concerns for chronic use.

6 Conclusion

Pulmonary delivery of therapeutic peptides has great potential to achieve the desired topical and systemic treatment via a non-invasive route with minimal risk of systemic side effects. The DPI system is theoretically preferable for the development of peptide delivery systems because of its higher stability than the liquid form, not only in storage conditions but also in physiological environments. However, there are still some challenges in the production conditions of peptide-loaded DPI and physiological barriers after pulmonary administration. Depending on the physicochemical properties of the target peptides, the powderization technique and its conditions should be carefully optimized, and suitable particle design strategies and excipients should be selected to maximize the stability and therapeutic potential. Although there are many approved excipients for oral and injection formulations, available excipients for DPI products are limited. However, the number of available excipient candidates will expand when the industry is willing to invest in exploring alternative excipients. In the future, the development of biologics, especially peptides and nucleic acids will gain more interest, and newer delivery systems may be developed with continued advances in particle design technologies.

References

1. Agu RU, Ugwoke MI, Armand M, Kinget R, Verbeke N. The lung as a route for systemic delivery of therapeutic proteins and peptides. Respir Res. 2001;2:198–209.
2. Ajmera A, Scherliess R. Stabilisation of proteins via mixtures of amino acids during spray drying. Int J Pharm. 2014;463:98–107.
3. Alabsi W, Acosta MF, Al-Obeidi FA, Hay M, Polt R, Mansour HM. Synthesis, physicochemical characterization, in vitro 2D/3D human cell culture, and in vitro aerosol dispersion performance of advanced spray dried and co-spray dried angiotensin (1–7) peptide and PNA5 with Trehalose as microparticles/nanoparticles for targeted respiratory delivery as dry powder inhalers. Pharmaceutics. 2021;13
4. Ambikanandan M. Challenges in delivery of therapeutic genomics and proteomics. London: Elsevier; 2011.
5. Amore E, Manca ML, Ferraro M, Valenti D, La Parola V, Di Vincenzo S, Gjomarkaj M, Giammona G, Bondi ML, Pace E. Salmeterol Xinafoate (SX) loaded into mucoadhesive solid lipid microparticles for COPD treatment. Int J Pharm. 2019;562:351–8.

6. Andrews JM, Roberts CJ. A Lumry-Eyring nucleated polymerization model of protein aggregation kinetics: 1. Aggregation with pre-equilibrated unfolding. J Phys Chem B. 2007;111:7897–913.
7. Atanasova KR, Reznikov LR. Neuropeptides in asthma, chronic obstructive pulmonary disease and cystic fibrosis. Respir Res. 2018;19:149.
8. Babenko M, Alany RG, Calabrese G, Kaialy W, Elshaer A. Development of drug alone and carrier-based GLP-1 dry powder inhaler formulations. Int J Pharm. 2022;617:121601.
9. Bailon P, Won CY. PEG-modified biopharmaceuticals. Expert Opin Drug Deliv. 2009;6:1–16.
10. Blankenship K. The top 20 drugs by global sales in 2019. Fierce Pharma; 2019.
11. Bosquillon C, Rouxhet PG, Ahimou F, Simon D, Culot C, Preat V, Vanbever R. Aerosolization properties, surface composition and physical state of spray-dried protein powders. J Control Release. 2004;99:357–67.
12. Carlson TL, Lock JY, Carrier RL. Engineering the mucus barrier. Annu Rev Biomed Eng. 2018;20:197–220.
13. Carvalho TC, Peters JI, Williams RO, 3RD. Influence of particle size on regional lung deposition—what evidence is there? Int J Pharm. 2011;406:1–10.
14. Chan HK, Clark AR, Feeley JC, Kuo MC, Lehrman SR, Pikal-Cleland K, Miller DP, Vehring R, Lechuga-Ballesteros, D. Physical stability of salmon calcitonin spray-dried powders for inhalation. J Pharm Sci. 2004;93:792–804.
15. Chang LL, Pikal MJ. Mechanisms of protein stabilization in the solid state. J Pharm Sci. 2009;98:2886–908.
16. Chellappan DK, Prasher P, Saravanan V, Vern Yee VS, Wen Chi WC, Wong JW, Wong JK, Wong JT, Wan W, Chellian J, Molugulu N, Prabu SL, Ibrahim R, Darmarajan T, Candasamy M, Singh PK, Mishra V, Shastri MD, Zacconi FC, Chakraborty A, Mehta M, Gupta PK, Dureja H, Gulati M, Singh SK, Gupta G, Jha NK, George Oliver BG, Dua K. Protein and peptide delivery to lungs by using advanced targeted drug delivery. Chem Biol Interact. 2022;351:109706.
17. Chow MY, Lam JK. Dry powder formulation of plasmid DNA and siRNA for inhalation. Curr Pharm Des. 2015;21:3854–66.
18. Cicerone MT, Pikal MJ, Qian KK. Stabilization of proteins in solid form. Adv Drug Deliv Rev. 2015;93:14–24.
19. Costantino HR, Andya JD, Nguyen PA, Dasovich N, Sweeney TD, Shire SJ, Hsu CC, Maa YF. Effect of mannitol crystallization on the stability and aerosol performance of a spray-dried pharmaceutical protein, recombinant humanized anti-IgE monoclonal antibody. J Pharm Sci. 1998;87:1406–11.
20. D'angelo I, Casciaro B, Miro A, Quaglia F, Mangoni ML, Ungaro F. Overcoming barriers in Pseudomonas aeruginosa lung infections: engineered nanoparticles for local delivery of a cationic antimicrobial peptide. Colloids Surf,B Biointerfaces. 2015;135:717–25.
21. Depreter F, Pilcer G, Amighi K. Inhaled proteins: challenges and perspectives. Int J Pharm. 2013;447:251–80.
22. Dey T, Kalita J, Weldon S, Taggart CC. Proteases and their inhibitors in chronic obstructive pulmonary disease. J Clin Med. 2018;7
23. Du Z, Guan YX, Yao SJ, Zhu ZQ. Supercritical fluid assisted atomization introduced by an enhanced mixer for micronization of lysozyme: particle morphology, size and protein stability. Int J Pharm. 2011;421:258–68.
24. Eedara BB, Alabsi W, Encinas-Basurto D, Polt R, Mansour HM. Spray-dried inhalable powder formulations of therapeutic proteins and peptides. AAPS PharmSciTech. 2021;22:185.
25. Eedara BB, Rangnekar B, Doyle C, Cavallaro A, Das SC. The influence of surface active l-leucine and 1,2-dipalmitoyl-sn-glycero-3-phosphatidylcholine (DPPC) in the improvement of aerosolization of pyrazinamide and moxifloxacin co-spray dried powders. Int J Pharm. 2018;542:72–81.
26. Faghihi H, Vatanara A, Najafabadi AR, Ramezani V, Gilani K. The use of amino acids to prepare physically and conformationally stable spray-dried IgG with enhanced aerosol performance. Int J Pharm. 2014;466:163–71.

27. Fahy JV, Dickey BF. Airway mucus function and dysfunction. N Engl J Med. 2010;363:2233–47.
28. Frohlich E, Salar-Behzadi S. Oral inhalation for delivery of proteins and peptides to the lungs. Eur J Pharm Biopharm. 2021;163:198–211.
29. Ghadiri M, Young PM, Traini D. Strategies to enhance drug absorption via nasal and pulmonary routes. Pharmaceutics. 2019;11
30. Groneberg DA, Witt C, Wagner U, Chung KF, Fischer A. Fundamentals of pulmonary drug delivery. Respir Med. 2003;97:382–7.
31. Hong DX, Yun YL, Guan YX, Yao SJ. Preparation of micrometric powders of parathyroid hormone (PTH1-34)-loaded chitosan oligosaccharide by supercritical fluid assisted atomization. Int J Pharm. 2018;545:389–94.
32. Johnson KA. Preparation of peptide and protein powders for inhalation. Adv Drug Deliv Rev. 1997;26:3–15.
33. Jong T, Li J, Morton DA, Zhou QT, Larson I. Investigation of the changes in Aerosolization behavior between the jet-milled and spray-dried Colistin powders through surface energy characterization. J Pharm Sci. 2016;105:1156–63.
34. Jyoti K, Katare OP, Kamboj A, Madan J. Protamine sulphate coated poly (lactide-co-glycolide) nanoparticles of MUC-1 peptide improved cellular uptake and cytokine release in mouse antigen presenting cells. J Microencapsul. 2020;37:566–76.
35. Kamerzell TJ, Esfandiary R, Joshi SB, Middaugh CR, Volkin DB. Protein-excipient interactions: mechanisms and biophysical characterization applied to protein formulation development. Adv Drug Deliv Rev. 2011;63:1118–59.
36. Karn PR, Cho W, Park HJ, Park JS, Hwang SJ. Characterization and stability studies of a novel liposomal cyclosporin A prepared using the supercritical fluid method: comparison with the modified conventional Bangham method. Int J Nanomedicine. 2013;8:365–77.
37. Kim YH, Sioutas C, Shing KS. Influence of stabilizers on the physicochemical characteristics of inhaled insulin powders produced by supercritical antisolvent process. Pharm Res. 2009;26:61–71.
38. Krebs MR, Domike KR, Donald AM. Protein aggregation: more than just fibrils. Biochem Soc Trans. 2009;37:682–6.
39. Kubczak M, Michlewska S, Bryszewska M, Aigner A, Ionov M. Nanoparticles for local delivery of siRNA in lung therapy. Adv Drug Deliv Rev. 2021;179:114038.
40. Lam XM, Yang JY, Cleland JL. Antioxidants for prevention of methionine oxidation in recombinant monoclonal antibody HER2. J Pharm Sci. 1997;86:1250–5.
41. Leleux J, Williams RO, 3RD. Recent advancements in mechanical reduction methods: particulate systems. Drug Dev Ind Pharm. 2014;40:289–300.
42. Leong EWX, Ge R. Lipid nanoparticles as delivery vehicles for inhaled therapeutics. Biomedicine. 2022;10
43. Maa YF, Hsu CC. Protein denaturation by combined effect of shear and air-liquid interface. Biotechnol Bioeng. 1997;54:503–12.
44. Maa YF, Nguyen PA, Hsu SW. Spray-drying of air-liquid interface sensitive recombinant human growth hormone. J Pharm Sci. 1998;87:152–9.
45. Maa YF, Nguyen PA, Sweeney T, Shire SJ, Hsu CC. Protein inhalation powders: spray drying vs spray freeze drying. Pharm Res. 1999;16:249–54.
46. Manca ML, Valenti D, Sales OD, Nacher A, Fadda AM, Manconi M. Fabrication of polyelectrolyte multilayered vesicles as inhalable dry powder for lung administration of rifampicin. Int J Pharm. 2014;472:102–9.
47. Manning MC, Chou DK, Murphy BM, Payne RW, Katayama DS. Stability of protein pharmaceuticals: an update. Pharm Res. 2010;27:544–75.
48. Matsson P, Kihlberg J. How big is too big for cell permeability? J Med Chem. 2017;60:1662–4.
49. Misaka S, Aoki Y, Karaki S, Kuwahara A, Mizumoto T, Onoue S, Yamada S. Inhalable powder formulation of a stabilized vasoactive intestinal peptide (VIP) derivative: anti-inflammatory effect in experimental asthmatic rats. Peptides. 2010;31:72–8.

50. Muralidharan P, Mallory E, Malapit M, Hayes D, JR. & Mansour, H. M. Inhalable PEGylated phospholipid Nanocarriers and PEGylated therapeutics for respiratory delivery as aerosolized colloidal dispersions and dry powder inhalers. Pharmaceutics. 2014;6:333–53.
51. Nikjoo D, Van Der Zwaan I, Brulls M, Tehler U, Frenning G. Hyaluronic acid hydrogels for controlled pulmonary drug delivery-A particle engineering approach. Pharmaceutics. 2021;13
52. Onoue S, Matsui T, Kato M, Mizumoto T, Liu B, Liu L, Karaki S, Kuwahara A, Yamada S. Chemical synthesis and formulation design of a PEGylated vasoactive intestinal peptide derivative with improved metabolic stability. Eur J Pharm Sci. 2013;49:382–9.
53. Onoue S, Misaka S, Ohmori Y, Sato H, Mizumoto T, Hirose M, Iwasa S, Yajima T, Yamada S. Physicochemical and pharmacological characterization of novel vasoactive intestinal peptide derivatives with improved stability. Eur J Pharm Biopharm. 2009;73:95–101.
54. Onoue S, Ohshima K, Debari K, Koh K, Shioda S, Iwasa S, Kashimoto K, Yajima T. Mishandling of the therapeutic peptide glucagon generates cytotoxic amyloidogenic fibrils. Pharm Res. 2004;21:1274–83.
55. Pangeni R, Meng T, Poudel S, Sharma D, Hutsell H, Ma J, Rubin BK, Longest W, Hindle M, Xu Q. Airway mucus in pulmonary diseases: Muco-adhesive and muco-penetrating particles to overcome the airway mucus barriers. Int J Pharm. 2023;634:122661.
56. Patel A, Cholkar K, Mitra AK. Recent developments in protein and peptide parenteral delivery approaches. Ther Deliv. 2014;5:337–65.
57. Patton JS, Byron PR. Inhaling medicines: delivering drugs to the body through the lungs. Nat Rev Drug Discov. 2007;6:67–74.
58. Poursina N, Vatanara A, Rouini MR, Gilani K, Najafabadi AR. The effect of excipients on the stability and aerosol performance of salmon calcitonin dry powder inhalers prepared via the spray freeze drying process. Acta Pharma. 2016;66:207–18.
59. Qiu Y, Man RCH, Liao Q, Kung KLK, Chow MYT, Lam JKW. Effective mRNA pulmonary delivery by dry powder formulation of PEGylated synthetic KL4 peptide. J Control Release. 2019;314:102–15.
60. Raman B, Chatani E, Kihara M, Ban T, Sakai M, Hasegawa K, Naiki H, Rao C, M. & Goto, Y. Critical balance of electrostatic and hydrophobic interactions is required for beta 2-microglobulin amyloid fibril growth and stability. Biochemistry. 2005;44:1288–99.
61. Ribeiro Dos Santos I, Richard J, Pech B, Thies C, Benoit JP. Microencapsulation of protein particles within lipids using a novel supercritical fluid process. Int J Pharm. 2002;242:69–78.
62. Sato H, Tabata A, Moritani T, Morinaga T, Mizumoto T, Seto Y, Onoue S. Design and characterizations of inhalable poly(lactic-co-glycolic acid) microspheres prepared by the fine droplet drying process for a sustained effect of Salmon calcitonin. Molecules. 2020;25
63. Sharma A, Vaghasiya K, Gupta P, Singh AK, Gupta UD, Verma RK. Dynamic mucus penetrating microspheres for efficient pulmonary delivery and enhanced efficacy of host defence peptide (HDP) in experimental tuberculosis. J Control Release. 2020;324:17–33.
64. Shen YB, Du Z, Tang C, Guan YX, Yao SJ. Formulation of insulin-loaded N-trimethyl chitosan microparticles with improved efficacy for inhalation by supercritical fluid assisted atomization. Int J Pharm. 2016;505:223–33.
65. Shiehzadeh F, Tafaghodi M. Dry powder form of polymeric nanoparticles for pulmonary drug delivery. Curr Pharm Des. 2016;22:2549–60.
66. Singh A, Van Den Mooter G. Spray drying formulation of amorphous solid dispersions. Adv Drug Deliv Rev. 2016;100:27–50.
67. Slomkowski S, Gosecki M. Progress in nanoparticulate systems for peptide, proteins and nucleic acid drug delivery. Curr Pharm Biotechnol. 2011;12:1823–39.
68. Suzuki H, Moritani T, Morinaga T, Seto Y, Sato H, Onoue S. Amorphous solid dispersion of cyclosporine A prepared with fine droplet drying process: physicochemical and pharmacokinetic characterization. Int J Pharm. 2017;519:213–9.
69. Taipaleenmaki E, Stadler B. Recent advancements in using polymers for intestinal Mucoadhesion and Mucopenetration. Macromol Biosci. 2020;20:e1900342.
70. Telko MJ, Hickey AJ. Dry powder inhaler formulation. Respir Care. 2005;50:1209–27.

71. Torosantucci R, Schoneich C, Jiskoot W. Oxidation of therapeutic proteins and peptides: structural and biological consequences. Pharm Res. 2014;31:541–53.
72. Weers JG, Miller DP. Formulation Design of dry Powders for inhalation. J Pharm Sci. 2015;104:3259–88.
73. Yang X, Ma JK, Malanga CJ, Rojanasakul Y. Characterization of proteolytic activities of pulmonary alveolar epithelium. Int J Pharm. 2000;195:93–101.
74. Yao JF, Yang H, Zhao YZ, Xue M. Metabolism of peptide drugs and strategies to improve their metabolic stability. Curr Drug Metab. 2018;19:892–901.
75. Yeo SD, Lim GB, Debendetti PG, Bernstein H. Formation of microparticulate protein powder using a supercritical fluid antisolvent. Biotechnol Bioeng. 1993;41:341–6.
76. Yu S, Pu X, Ahmed MU, Yu HH, Mutukuri TT, Li J, Zhou QT. Spray-freeze-dried inhalable composite microparticles containing nanoparticles of combinational drugs for potential treatment of lung infections caused by Pseudomonas aeruginosa. Int J Pharm. 2021;610:121160.
77. Zapadka KL, Becher FJ, Gomes Dos Santos AL, Jackson SE. Factors affecting the physical stability (aggregation) of peptide therapeutics. Interface Focus. 2017;7:20170030.
78. Zhang Y, Mackenzie B, Koleng JJ, Maier E, Warnken ZN, Williams RO, 3RD. Development of an excipient-free peptide dry powder inhalation for the treatment of pulmonary fibrosis. Mol Pharm. 2020;17:632–44.
79. Zheng JY, Huang SS, Huang SH, Ye JJ. Colistin for pneumonia involving multidrug-resistant Acinetobacter calcoaceticus-Acinetobacter baumannii complex. J Microbiol Immunol Infect. 2020;53:854–65.
80. Zhou S, Zhang B, Sturm E, Teagarden DL, Schoneich C, Kolhe P, Lewis LM, Muralidhara BK, Singh SK. Comparative evaluation of disodium edetate and diethylenetriaminepentaacetic acid as iron chelators to prevent metal-catalyzed destabilization of a therapeutic monoclonal antibody. J Pharm Sci. 2010;99:4239–50.

Pulmonary Delivery of Antibody for the Treatment of Respiratory Diseases

Thomas Sécher and Nathalie Heuzé-Vourc'h

Abstract Over the past 30 years, therapeutic antibodies (Abs) have offered groundbreaking solutions for a wide range of diseases, including respiratory diseases, which represent a significant burden worldwide. The Ab market is continuously growing, with dozens of new Abs reaching clinical trials every month. While clinically approved Abs confirmed their potential as innovant therapeutics, preclinical studies showed that their efficacy may be bolstered by delivering the molecules locally. In fact, alternative delivery methods, addressing Abs to the disease site, have emerged and progressed to the clinic. Oral inhalation is the gold standard route for small molecules commonly used for the treatment of respiratory infections and inflammatory diseases (asthma, chronic obstructive pulmonary diseases (COPD)). It is also a thriving focus of research for Abs against respiratory diseases. This chapter proposes an overview of Abs delivered by inhalation, focusing mostly on liquid aerosols delivered to the lungs by nebulization. It describes Ab features, host biological properties and technical/scientific issues, which are important to consider for the development of inhaled Abs.

Keywords Inhalation · Biological barriers · Lungs · Mucus · Therapeutic antibody · PK/PD

1 Introduction

Respiratory diseases account for the most common causes of severe illness and death worldwide: lung infections, lung cancers and chronic obstructive pulmonary diseases (COPD) are among the top 10 major killers, causing one-sixth of all deaths,

T. Sécher (✉) · N. Heuzé-Vourc'h (✉)
INSERM U1100, Centre d'Etude des Pathologies Respiratoires, Tours, France

Université de Tours, Tours, France
e-mail: thomas.secher@inserm.fr; nathalie.vourch@med.univ-tours.fr

J. K. W. Lam, P. C. L. Kwok (eds.), *Respiratory Delivery of Biologics, Nucleic Acids, and Vaccines*, AAPS Introductions in the Pharmaceutical Sciences 8,
https://doi.org/10.1007/978-3-031-47567-2_2

and millions of people suffer from chronic respiratory diseases such as asthma and pulmonary hypertension. Overall, respiratory diseases enforce a huge health and economic burden: premature mortality, costs related to primary/hospital care, treatments, loss of productivity and disability-adjusted life-years (DALYs) lost [1]. Accordingly, Abs, which has proven successful to prevent/treat different pathological conditions, have a tremendous opportunity to benefit to patients with respiratory diseases. We and others have published several reviews on this topic and refer to them here if reader would like more details [2–5].

Briefly, several Abs are already approved to treat respiratory diseases, for non-small cell lung cancer, asthma and respiratory tract infections, as illustrated in Table 1. The first Ab approved for a respiratory condition, was pavilizumab (1998) which binds to the fusion (F) protein of human respiratory syncytial virus (RSV) and is used to prevent RSV infections in high-risk preterm neonates. All these Abs are delivered by the intravenous and subcutaneous routes, or intramuscularly. Pulmonary delivery of Abs has been a thriving focus of research for many Abs to treat respiratory diseases (Table 2), but there is no inhaled Ab approved yet.

Abs are glycoproteins belonging to the immunoglobulin superfamily. Most abs approved or in review in EU or US are of the IgG subclass (https://www.antibodysociety.org/resources/approved-antibodies/). As illustrated in Fig. 1, typical structure of IgG consists of four peptide chains – two identical κ or λ light chains and two γ heavy chains– connected by disulphide bonds and electrochemical interactions, reaching an approximately 150,000 Da molecular weight [6]. The fragment antigen binding (Fab) contains the complementarity determining regions (CDR) binding to the epitope on the target antigen. Monoclonal abs recognized only one epitope, and IgGs may bind to two epitopes simultaneously, since they comprise two Fab regions (Fig. 1). The fragment crystalline (Fc) is located on the heavy chains and is responsible for the effector functions of the Ab, through binding to either Fcγ receptors on immune cells (natural killer cells, macrophages, etc) or complement cascade enzymes (Fig. 2). Among IgG subclass, it is admitted that IgG1 and IgG3 display more potent effector mechanisms, as compared to IgG2 and IgG4, as they bind with different affinity to Fcγ receptors [7]. IgG1 and IgG3 also efficiently activate the classical route of complement, while IgG4 have limited complement activation and only under specific conditions for IgG2. Finally, the Fc region also contains a highly conserved N-glycosylation site and the binding site to neonatal Fc receptor (FcRn), which is important for Ab pharmacokinetics (PK). Engineering Ab to sequentially replace murine sequence-derived amino acids with human ones (Fig. 1) has been done to significantly reduce immunogenicity [8, 9]. However, it remains unclear whether fully human Abs is less risky for immunogenicity than the humanized constructs. Some humanized and full-human Ab still carry immunogenicity risk.

Table 1 Approved antibodies for respiratory diseases

Indication	Generic name (Trade name)	Sponsoring company	Target	Antibody format	Date of approval
NSCLC	Atezolizumab (Tecentriq™)	Roche	PDL1	Humanized IgG1	2016
	Necitumumab (Portrazza™)	Eli Lilly	EGFR	Human IgG1	2016
	Nivolumab (Opdivo™)	BMS	PD1	Human IgG4	2015–2016
	Pembrolizumab (Keytruda™)	Merck	PD1	Humanized IgG4	2015
	Ramucirumab (Cyramza™)	Eli Lilly	VEGFR	Human IgG1	2015
	Racotumomab (Vaxira™)	Recombio	Ganglioside mimical	Murine IgG1	2013[a]
	Bevacizumab (Avastin™)	Sanofi/Genentech	VEGF	Humanized IgG1	2007
	Ipilimumab (Yervoy™)	BMS	CTLA4	Human IgG1	2020
	Durvalumab (Imfinzi™)	AstraZeneca	PD1	Human IgG1	2018
Asthma	Benralizumab (Fasenra™)	MedImmune	IL5R	Humanized IgG1	2017
	Reslizumab (Cinquil™)	TEVA	IL5	Humanized IgG4	2016
	Omalizumab (Xolair™)	Novartis	IgE	Humanized IgG1	2015–2016
	Mepolizumab (Nucala™)	GSK	IL5	Humanized IgG1	2015
	Tezepelumab (Tezspire™)	Amgen	TSLP	Human IgG2	2021
SARS-CoV-2 infection	Sotrovimab (Xevudy™)	GSK/Vir Biotechnology	Spike protein	Human IgG1	2021
	Regdanvimab (Regkirona™)	Celltrion/Inhalon Biopharma	Spike protein	Human IgG1	2021
	REGEN-COV (Ronapreve™)	Regeneron	Spike protein	Human IgG1	2021
	Bamlanivimab + etesevimab	Eli Lilly	Spike protein	Human IgG1	2021
	Amubarvimab + romlusevimab	Brii Biosciences	Spike protein	Human IgG1	2021[b]
	Tixagevimab + cilgavimab (Evusheld™)	AstraZeneca	Spike protein	Human IgG1	2022[c]

(continued)

Table 1 (continued)

Indication	Generic name (Trade name)	Sponsoring company	Target	Antibody format	Date of approval
ARDS	Levilimab (Ilsira™)	BIOCAD	IL-6R	Human IgG1	2021[d]
	Olokizumab (Artlegia ™)	UCB	IL-6	Humanized IgG4	2020[d]
	Sarilumab (Kevzara™)	Sanofi	IL-6R	Human IgG1	2021
	Tocilizumab (Actemra™)	Roche	IL-6R	Human IgG1	2021
Respiratory syncitial virus infection	Palivizumab (Synagis™)	MedImmune	F-protein	Humanized IgG1	1998
	Nirsevimab (Beyfortus™)	AstraZeneca	F-protein	Human IgG1	2023
Pulmonary anthrax	Obiltoxaximab (Anthim™)	Elusys Therapeutics	PA-antigen	Chimeric IgG1	2016
	Raxibacumab (Abthrax™)	GSK	PA-antigen	Human IgG1	2013
Nasal polyposis	Dupilumab (Dupixent™)	Sanofi	IL-4R	Human IgG4	2019

ARDS, Acute respiratory distress syndrome; CTLA4, cytotoxic T-lymphocyte antigen 4, *EGFR,* epidermal growth factor receptor; *EOS,* eosinophil in blood; *NSCLC,* non-small-cell lung cancer; *PDL1,* programmed cell death ligand 1; *VEGF,* vascular endothelial growth factor; *VEGFR,* vascular endothelial growth factor receptor

[a]in Argentina and Cuba

[b]in China

[c]in Russia

[d] in European Union

To date, most Abs on the market are of the IgG1 subclass, but both IgG4 and IgG2, with different functional activity, have also been approved. The preference for IgG1 may be explained by the strong effector functions and the longer half-life of this subclass, associated with the fact that oncology is a major application field for Ab. Ab for respiratory diseases are monoclonal and IgG1, apart from reslizumab, nivolumab and pembrolizumab, which are IgG4 (Table 1). It is noteworthy that Ab fragments and mimetics, with artificial frameworks (anticalins, DARPin, affibody, etc.) are also of interest, for pulmonary delivery (Table 2). They can offer different PK-PD behaviour, functionality, immunogenicity, safety or avoid intellectual property issues [10–13]. They offer the advantage of being produced easily, resulting in faster and higher bioproduction rates/yields and reduced costs. Because they often lack the Fc domain, they are deprived of effector functions and are cleared faster through renal excretion unless they are conjugated to albumin or pegylated to extend their half-life. Several Ab fragments such as abciximab, a chimeric IgG1 Fab raised against GPIIb/IIIa used to prevent blood clots during angioplasty, already reached

Table 2 Inhaled antibodies for respiratory diseases

Indication	Name	Sponsoring company	Target	Antibody format	Development stage	ClinicalTrial.gov identifier[a]
Asthma/COPD	Omalizumab	Novartis	IgE	Human IgG	Discontinued in Phase III	https://doi.org/10.1164/ajrccm.155.6.9196082 , https://doi.org/10.1164/ajrccm.160.3.9810012
	LQ036	Shanghai Novamab Biopharmaceuticals	unknown	Single-domain Ab	Phase I	NCT04993443
	VR942	Vectura	IL-13	Humanized $F(ab')_2$	Discontinued after phase I	NCT02473939
	CSJ-117	Novartis	TSLP	Human Fab	Phase II	NCT04882124
COVID-19	CT-P63 + CT-P66	Celltrion	Spike protein	Human IgG	Phase III	NCT05224856
	IBIO-123	Immune Bioscience	Spike protein	Human IgG	Phase I	NCT05298813
	IGM-6268	IGM Biosciences	Spike protein	Human IgM	Phase I	NCT05184218
	DZIF-10c (BI 767551)	University of Cologne/Boehringer Ingelheim	Spike protein	Human IgG	Discontinued after phase I/II	NCT04631705/NCT04631666
Respiratory syncitial virus infection	ALX-0171	Ablynx/Sanofi	F-protein	Single-domain Ab	Discontinued after Phase II	NCT03418571
Acute lung injury	GSK2862277	GSK	TNF-R1	Single-domain Ab	Discontinued after Phase II	NCT02221037

IgE, Immunoglobulin E; *IL*, interleukin; *TNF-R1*, tumour necrosis factor receptor 1; *TSLP*, thymic stromal lymphopoietin
[a]NCT number of the latest on-going clinical trials are indicated

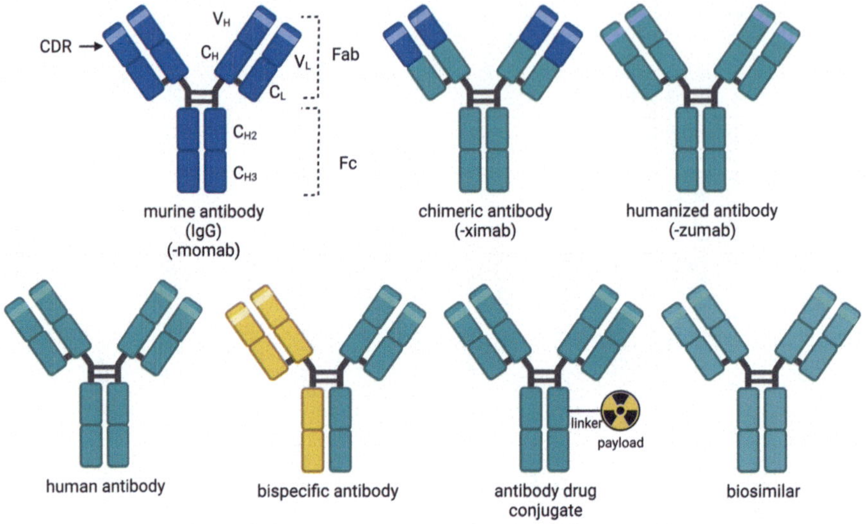

Fig. 1 Structure and format of antibody-based therapeutics. The antibody international non-proprietary names (INN) aims to provide clear identification of antibodies. The "-mab" stem indicate monoclonal antibody-based therapeutics. Mouse monoclonal antibody are indicated by the suffix "-mo-", such as muromomab, a mouse monoclonal antibody raised against CD3. The suffix -xi- indicates chimeric antibodies, which comprise murine variable regions conjugated to human constant regions, resulting in a molecule which is about 65% human. The suffix -zu- indicates humanized antibodies containing murine CDR grafted in human antibodies and resulting in an Ab which is approximately 95% human. Finally, the suffix -u- corresponds to fully human Abs containing 100% sequence derived from human genetic repertoire and are obtained historically from transgenic animals or screening of Ab libraries derived from human B-cell repertoire [164]. Bispecific antibody comprises two different mAbs that binds to two different types of antigen. Antibody-drug conjugate (ADC) is linked, through enzymatic or chemical reactions, to a payload, with specific pharmacological properties. Biosimilar antibody is a "generic" version of the reference human Ab. CDR, complementary-determining region, C_H, heavy chain constant domain; C_L, light chain constant domain; V_H, heavy chain variable domain; V_L, light chain variable domain; Fab, fragment antigen-binding; Fc, fragment crystallizable

regulatory approval (Table 1). Similarly, the mimetics DX-88, a kunitz domain binding to plasma kallikrein, has been approved by the Food and Drug Administration (FDA), in 2012, in hereditary angioedema.

The ab field is continuously evolving and products, approved or in development, may be univalent, divalent (bispecific Ab) or multivalent, with mutations to silence or enhance Fc-effector functions [10, 11], half-life extensions or conjugated to drug/radioisotope (Ab drug conjugate (ADC)) to improve PK or enhance pharmacodynamic (PD) properties (Fig. 1). Finally, it is noteworthy that biosimilars of Abs, which correspond to a "generic" version of the reference Ab (innovator/originator) with the same amino acid sequence but produced from different cells and

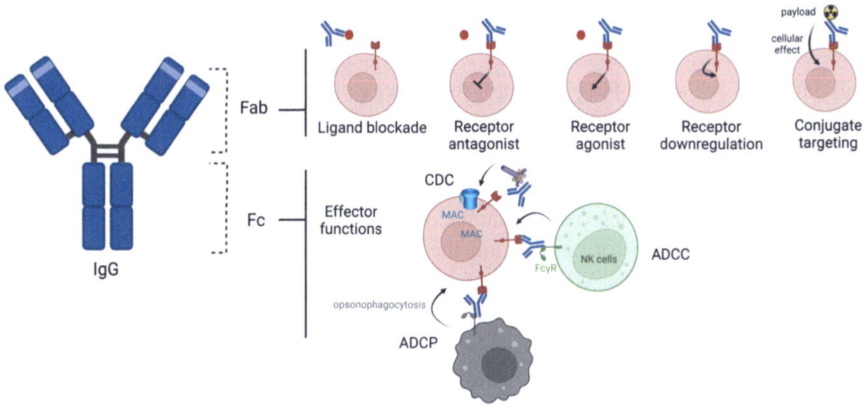

Fig. 2 Multiple modes of action of full-length IgG. The binding of the Ab to its epitope, through its Fab fragment, can result in: (1) soluble ligand blockade, thereby preventing them to activate their cognate receptor, (2) blocking/activating receptor membrane function, preventing or mimicking ligand binding to their receptor and subsequent blocking/activation of signal transduction, (3) receptor internalization and downregulation, (4) targeted delivery of payload drug (radioisotope, cytotoxic agent, antibiotic or cytokine) to specific cells, for antibody-drug conjugate (ADCs). Here, the ADC is presented with a payload drug corresponding to a radioisotope [2]. Using the specificity of the antibody to its target will precisely deliver payload (radio/chemotherapy agent, antibiotic, cytokine) to target cell. IgGs may trigger different types of effector functions: antibody-dependent cellular cytotoxicity (ADCC), and antibody-dependent cellular phagocytosis (ADCP), which have been shown to be crucial for the therapeutic efficacy of many antibodies. IgGs can also activate the complement enzyme cascade, initiated through binding of C1q to the Fc fragment and subsequent activation of C1q. This ultimately leads to deposition of C3b to further opsonize the target and the formation of the membrane attack complex, C5–C9, triggering disruption of the bilipid membrane (and formation of a membrane attack complex -MAC). ADC, Antibody-drug conjugate; FcγR, Fcγ receptor

manufacturing processes, are a thriving focus of development, as patents of reference Ab expire, to reduce medical expenditures and gain new markets (Fig. 1). Ab biosimilars have the same formulation to treat the same disease and have to demonstrate pharmacological comparability to the reference. However, biosimilars and generic drugs are different, as generics contain an identical chemically active ingredient to their reference products, while biosimilars are very similar but not identical, as they are naturally variable to the reference ab [14].

There are currently no FDA/European Medicine Agency (EMA)-approved inhaled Ab products. This may be explained by several factors, which are detailed in the following sections and relate to the stability of Ab during aerosolization, PK and PD considerations. This chapter also reports the different Ab-based therapeutics that have been studied by inhalation (Table 2) and attempts to explain why pulmonary delivery of Ab has not materialized in clinical success yet.

2 Stability of Antibodies During Aerosolization

Inhalation of Ab in the lungs through the airways is conditioned by the generation of an aerosol containing particles between 1 and 5 μm, which will deposit in the different parts of the respiratory tract. The region of the lower respiratory tract in which aerosol particles will deposit in the lungs is dictated by the physicochemical properties of the particles (i.e. geometric size, density, and shape), which depends on both the inhaler performances and the drug formulation. There are three main types of inhalers: dry powder inhalers (DPI), pressurized metered dose inhalers (pMDI) and nebulizers. In this section, we will focus on nebulizers to deliver Abs as liquid aerosols.

Nebulizers (jet, ultrasonic and mesh) accommodate liquid solution or suspension and are often used as a first step in the development process of inhalation-based administration of protein therapeutics. Indeed, nebulizers avoid the drying steps of DPI, are suitable for all clinical situations, accommodate large volumes with less pressure on having formulations with high concentrations of Ab, and often enable greater pulmonary deposition than DPI. Moreover, nebulized formulations are less expensive to produce and assess. However, liquid formulations may be less stable for prolonged storage, potentially resulting in Ab degradation, and nebulization is associated with longer administration time.

The following sections explore the impact of aerosolization on Ab, the importance of the device and the formulation to ensure Ab stability and the consequences of the instability of Ab during aerosolization.

2.1 Aggregation as a Marker of Ab Stability

Due to their labile molecular structures, Abs are susceptible to various stresses involved in the generation of aerosol particles/droplets, leading to physical degradation of the protein including denaturation and, in the end, promoting aggregation [15]. Aggregates are formed by the assembly of native and/or unfolded Ab by weak interactions (Van der Waals interactions, hydrogen bonding, hydrophobic and electrostatic interactions) known as physical aggregation or self-association or by covalent bonding leading to covalent aggregation. Both physical and covalent aggregation may result in soluble and insoluble aggregates. Insoluble aggregates may consist solely of Ab or contain contaminants, excipients, etc. Although not reported yet during the aerosolization process, Ab is also susceptible to chemical degradations (deamidation, oxidation and fragmentation), which typically occur through interaction with certain excipients. Chemical degradations are intertwined with physical degradation, as they may ultimately lead to (covalent) aggregation [15, 16]. Aggregation is a major marker of Ab instability during nebulization, but it is not known whether aggregates resulted from chemical and/or physical degradations.

Different intrinsic factors of the Ab may affect their stability and influence their propensity to aggregate upon stresses: their primary and tertiary structure, their subclass, or the isoelectric point of their CDR [15]. External factors may also impact Ab aggregation, such as Ab concentration, temperature variations, interfaces, light, excipient, agitation, and shearing [15]. Ab aggregation may be monitored by different orthogonal methods (dynamic light scattering, size exclusion chromatography, etc.) enabling the characterize of aggregates based on their size or the order of Ab assemblies [17–19]. It is noteworthy that analysing aggregation following nebulization requires collecting/condensating the aerosol back into a solution to implement analytical methods. Recently, we showed that the collection device interferes with Ab stability, inducing a bias in aggregation monitoring, which should be considered during inhaled Ab development [20]. According to the European pharmacopoeia, a high-quality Ab product should be free from visible aggregates (or particles) and contain a limited quantity of particles above 10 and 25 μm [21]. These recommendations apply to parenteral Ab products, but there is no guideline for inhaled Ab, yet.

2.2 *Importance of the Device*

To generate a liquid Ab aerosol by nebulization, obtaining aerosol particles with an appropriate aerodynamic size for lung deposition is important, while preventing protein aggregation.

Nebulization exposes Ab to a huge air-liquid interface (24–1500 m^2) where the protein has a tendency to adsorb and denature and may be, depending on the device, associated with temperature rise and Ab recirculation. Among the three types of nebulizers (jet, ultrasonic and mesh-nebulizers), mesh-nebulizers have been shown to be less deleterious for Ab, with fewer aggregates generated upon nebulization as compared to ultrasonic and/or jet nebulizers [22]. Although aggregation is the most visible and reported manifestation of Ab instability during nebulization, it is possible that chemical degradation of Ab – in particular oxidation, may occur during nebulization, as observed for other protein therapeutics [23]. Interestingly, several Ab developed for mesh-nebulization have reached clinical trials and have been reported to be safe. GSK 1995057, a single-domain anti-TNF receptor 1 antibody was developed for mesh-nebulization with PARI eFLOW® to limit lung acute injury following oesophagectomy surgery. ALX-0171 nanobody, against the RSV, was first tested with the Aerogen solo® mesh-nebulizer for use in adults and next delivered with the FOX®-flamingo mesh-nebulizer in a phase 2b trial in hospitalized children [22] (Table 2). Finally, several anti-SARS-CoV-2 Abs are currently in clinical trials given by mesh-nebulization [24].

It is noteworthy that each Ab has a different vulnerability to stress and not all devices in the same category are equivalent, potentially being deleterious on one specific Ab [25–27]. Thus, the device should be selected carefully, considering its performances and Ab formulation stability and following a specific development approach.

2.3 Importance of the Formulation

As already mentioned, Abs are highly susceptible to stress that may occur at different stages, from Ab bioproduction up to administration to patients, including during aerosolization. Accordingly, scientists develop adequate formulation to ensure shelf-life stability and appropriate quality of Ab product. Recently, we showed that liquid formulations developed for intravenous injection could not be easily repurposed for inhalation by nebulization. Indeed, the stresses applied during the pharmaceutical development of parenteral Abs, such as shaking, temperature changes, do not recapitulate the ones of nebulization [26].

Inhaled Ab formulations rely on preventing protein aggregation and degradation in liquids to ensure lung tolerance. As previously demonstrated for other inhaled proteins and Abs delivered parenterally, the addition of excipients and selection of the appropriate buffering system in formulation help to preserve protein structure and function during stressing processes [23, 28–30]. There are many excipients Generally Recognized As Safe (GRAS list), to date, but only few of them are approved for pulmonary delivery due to the lack of toxicological studies for inhaled excipients. Thus, the list of excipients to stabilize Abs for pulmonary delivery is limited and adding a new excipient in a formulation must be considered cautiously since it will account for extra workload, time, cost, and potential regulatory delays/rejection.

In liquid formulations, stabilizers include buffering or pH-adjusting agents, salts, and surfactants. The buffering system is expected to maintain the pH of the formulation. The right type of buffering agent, its concentration and pH should be chosen appropriately since they may influence the propensity of Ab to aggregate. For example, the buffering agent and pH were shown to limit aggregation and deamidation during the drying of an anti-IL-13 fragment [23]. Similarly, we observed that the selection of the buffering system was critical to maintaining IgG1 stability during mesh-nebulization (unpublished). It is noteworthy that the formulation of ALX-0171, a trimeric nanobody, comprised only NaCl as an osmolality agent and phosphate as buffer component in addition to the active principal ingredient (50 mg/mL), which was sufficient to ensure the stability of the Ab during mesh-nebulization [22]. Surfactants are often used to prevent Ab adsorption at interfaces (air-liquid, liquid-solid, etc.) and have proven successful in stabilizing IgG during mesh-nebulization [26, 31, 32]. Salts are mostly used to adjust the osmolarity to app. 300 mOsmol/L to favour formulation tolerance after airway delivery and may change the stability of Ab by altering the surrounding electrostatic environment [30]. If reader like to understand more in-depth the role and impact of different excipients on Ab stability, we recommend the recent review from Le Basle et al. [15].

To summarize, development of an inhaled Ab drug product is intertwined with the device selection and formulation components, which must be chosen to ensure Ab physical/chemical stability, for optimal aerosol performances and particle deposition in the respiratory tract, and avoid lung toxicity.

2.4 Consequences of Antibody Instability During Aerosolization

The impact of chemical and physical degradations on Ab may highly depend on their location and the nature of the aggregates produced [15]. For instance, chemical degradation in the Fc fragment may interfere with Fc-effector functions and interactions with FcRn, thereby modifying Ab PK and PD. Chemical degradation occurring in the CDR may impair Ab binding to its target antigen and loss of potency. Aggregation may modify inhaled Ab potency, increase or loss, as recently reported [20].

The lungs are a mucosa, sentinelled by a high density of immune cells, which may recognize aggregated Ab as antigens and produce antidrug antibody (ADA). The clinical manifestations of ADA range from no noticeable effect and changes in Ab PK-PD to hypersensitivity reactions. Hypersensitivity reactions may vary from mild to anaphylaxis. After Ab inhalation, the generation of ADA in preclinical models is heterogeneous and specific to the animal species and the molecule. ADA may be associated with alteration of PK and concomitantly accompanied by infiltration of immune cells into the lungs and hypercellularity in bronchus-associated lymphoid tissue, but the correlation of those lung pathology findings with ADA is uncertain [33, 34]. It is well-accepted that ADA in preclinical species do not predict immunogenicity in humans, and the high incidence and level of ADA in preclinical models usually do not prevent inhaled Ab from progressing into clinical trials [34]. It is noteworthy that the pulmonary route, together with the intradermal and subcutaneous routes, is considered more immunogenic than the intravenous and oral routes [35]. However, the literature on this topic remains limited and inconclusive.

The likelihood of inhaled Ab to produce ADA depends on multiple factors: (1) the ab itself such as the presence and percentage of non-host species sequences, its mechanism of action, internalization upon binding to its antigen, post-translational modifications and aggregation [36], (2) the dose and regimen of inhaled Ab and (3) patient-specific factors (disease state, concurrent medication, etc). Aggregation is a major factor promoting immunogenicity [37] and ADA generation, but it may be associated to other deleterious immune effect, as recently reported [27]. Indeed, the production of Ab aggregates during mesh-nebulization resulted in a profound and sustained local and systemic depletion of immune cells after delivery through the pulmonary route, which was attributable to cell death. This immunocytotoxic effect was dependent on the route of administration of aggregates and their amounts [27].

Thus, controlling physical and chemical degradations of inhaled Ab is critical to minimize risks for patients.

3 PK of Inhaled Ab

Ab PK describes the dynamic fate of an Ab in a living animal and is characterized by Ab absorption, distribution, metabolism, and elimination [38]. Ab PK will affect the magnitude and duration of the response. Usually, PK parameters are estimated by measuring drug concentrations in the systemic circulation from which the behaviour of the drug at the target site is extrapolated from compartmental models. In the context of Ab inhalation, PK evaluation is challenging due to sampling methods which are technically, or ethically questionable, limited understanding of molecular processes involved in lung absorption. As described below and due to their high molecular weight, Abs does not passively diffuse through the different compartments of the body from the systemic compartment. In addition, several pulmonary diseases including cancer or fibrotic diseases are associated with marked limitation of blood supply [39–41]. Thus, the targeted delivery of Ab to the lungs using the systemic route will result in a very low drug concentration, legitimating inhalation as a more relevant route of administration for the treatment of respiratory diseases [42–45]. This mutual hindrance makes inhaled Ab to pass poorly from the airways into the systemic circulation; their concentration in the lungs is expected to be higher than in the blood [46–48]. Consequently, the PK profile of Ab in the systemic compartment cannot easily be extrapolated to inhaled Ab, which PK, according to the complexity of the deposition and absorption mechanisms described thereafter, is challenging to evaluate.

3.1 Fate of Abs After They Deposit into the Pulmonary Tract

Given the anatomic characteristics of the lungs and their modifications during diseases [52], it is important to achieve an appropriate deposition pattern of the Ab depending on the expression of its target antigen. As for all inhaled drugs, the particle aerodynamic diameter of Ab aerosols is one of the most critical parameters that dictate aerosol performance and deposition into the lungs [30, 53–55]. Besides aerodynamic considerations, other parameters described thereafter may influence Ab fate in the lungs.

3.1.1 Lung Absorption

The successful landing of the particles on the airway surface is not necessarily associated with pulmonary absorption as it will be influenced by particle intrinsic features, clearance processes and lung barriers. The absorption of biologics at the respiratory interface involves highly complex mechanisms, and for some of them are not yet well characterized. However, it appears that the rate of absorption is mainly dependent on the size of the inhaled biologics [49–51] with the half-time of alveolar absorption increased proportionally with the molecular mass of the inhaled biologics [52]. Small peptides/proteins with a molecular weight below 40 kDa passed quickly in the bloodstream [53–55] while larger molecules, like full-length Ab [47, 56] or ab fragments [57, 58] exhibited limited bioavailability. Other parameters affecting the absorption will include pH, electrical charge, surface activity and solubility/stability of the drug in the pulmonary environment (reviewed in [48]). Biologics administered in the airways can be absorbed by three distinct mechanisms: paracellular diffusion via tight junctions, transcellular diffusion via vesicular endocytosis or pinocytosis, and receptor-dependent transcytosis [59, 60]. Low molecular weight inhaled biologics will be absorbed preferentially by the paracellular route, while larger molecules seem to exploit transcellular passage [61]. Small peptides can be absorbed by receptor-mediated transcytosis using the peptide transporters [62] while immunoglobulin uses a combination of pinocytosis with receptor-mediated transcytosis, using the FcRn, FcγR or through Fab-target binding [63, 64].

3.1.2 Lung Exposure

As compared to the half-life of systemic IgG lasting for ~18–21 days, inhaled Abs are quickly eliminated from the lungs, within 1–2 days after administration. In steady-state conditions, this exposition appears to be non-linear, biphasic, with a continuous disappearance of the Ab in the airway compartment for ~24 h and a limited and moderate passage into the systemic compartment [46, 47, 56, 65]. The first phase may be explained by distribution to the systemic compartment and/or attributed to the mechanisms accounting for Ab elimination from the airways, and that mainly include (i) exo/endogenous catabolism or (ii) target-mediated drug disposition (TMDD). Ab fragments derived from IgGs but lacking the Fc domain have a shorter half-life after intravenous injection, though it is not clear for inhalation. They also diffuse better in the different compartments or within solid tumours due to their smaller molecular size.

3.2 Lung Clearance

Even if an inhaled particle has successfully landed on the mucosal surface, complex physiological structures and mechanisms of the lungs may limit therapeutic efficacy. Indeed, inhaled particles must overcome mucus/surfactant entrapment, mucociliary clearance, degradation by lung proteases and phagocytosis by immune cells before interacting with the epithelial barrier. It is noteworthy that several pathological conditions, including COPD, cystic fibrosis (CF), asthma or infections, display modified barriers, which may substantially affect the pulmonary delivery of therapeutics.

3.2.1 Mucociliary Barrier

Mucus is a viscoelastic hydrogel coating the mucosal surface of the upper and central lung. It is continuously produced by secretory cells, including Goblet cells, club cells and other submucosal glands, and acts as a lubricant, maintaining moisture above the epithelial cells [66]. Mucus acts also as the primary barrier to billions of pathogens, allergens, irritants and dust that are inhaled daily [67]. Mucus is a gel-like structure mainly composed of water (>95%), proteins, which mucins represent the largest family, DNA, lipids, electrolytes, and cellular debris [66, 68]. Mucins are glycoproteins with periodic carboxy- and amino-terminal domains promoting cross-linking that confer to the mucus a network structure with a viscous nature [69]. The three-dimensional mesh structure of mucus, generate pores with sizes ranging ~10–100 nm and thickness to several micrometres (thicker in the trachea than in the bronchi) [70] enabling mucus to entrap inhaled drug particle and prevent their penetration to lung epithelium [71, 72]. Even at steady-state, large therapeutic protein like immunoglobulins exhibited reduced diffusion coefficient in mucus [73, 74]. Entrapped particles in the mucus will be moved toward the pharynx/larynx, thanks to the coordinated beating of cilia lining the upper airways where it will be swallowed, such a process constituting the mucociliary clearance. Intratracheal delivery to the lung revealed that a substantial fraction (~30%) of anti-IL17 Ab fragment was removed by mucociliary clearance [58]. In fact, Abs are highly charged and hydrophilic molecules that will tightly interact with mucus components limiting drug absorption [75]. Interestingly, it has been reported that the Fc fragment of Ab is the moiety exhibiting the highest electrostatic interactions, due to its negatively charged residues, with mucin fibres, controlling diffusion rate in the mucus. In fact, multimeric immunoglobulins (IgA, IgM) displayed limited diffusion in the mucus [73].

In many lung diseases, the mucus is severely altered. For example, asthma, COPD and CF share the symptom of mucus hypersecretion, resulting in a thicker mucus layer reaching >250 μm in CF [76]. In addition, the rheological properties of mucus with increased viscosity and rigidity due to the increase of cross-linking between mucins will stiffen the mucus layer [77]. COPD and CF are also characterized by mucus dehydration and shrinkage of pore size estimated to be smaller than 100 nm [78]. In fact, if the epithelial surface of highly dehydrated, the osmotic modulus of the mucus layer will be dramatically reduced and it will constrict the peri-ciliary layer, eventually stopping the mucociliary clearance, as observed during CF [79]. All these abnormalities may impair Ab penetration and diffusion toward the lung epithelium, as previously described for other therapeutics [80–82].

3.2.2 Surfactant Barrier

The anatomical structure of the alveoli is substantially different from that of the central lung. In fact, the epithelial surface is only composed of type I and II pneumocytes, covered with a thin layer of surfactant. This aqueous fluid is composed of phospholipids (~90%) and surfactant proteins (~10%), including hydrophilic proteins SP-A and SP-D, and hydrophobic proteins SP-B and SP-C [83, 84]. Its primary function is to lower air-liquid interface tension to prevent alveoli collapse during ventilation. However, large proteins like Ab may interact with surfactant components promoting the formation of a corona around the inhaled particle [85]. This may trigger aggregation and subsequent removal of therapeutic drug. This clearance process is mainly mediated by alveolar macrophages. In the peripheral lungs, these phagocytes play a significant role in pulmonary clearance processes by internalizing and degrading inhaled particles with size ranging from 0.5 to 5 μm. Phagocytosis by alveolar macrophages may become significant for protein with molecular weight >40 kDa [86]. Interestingly, inhaled proteins that have interact with hydrophilic surfactant proteins SP-A and SP-D are more prone to phagocytosis by alveolar macrophages [87]. Confocal imaging studies have revealed that alveolar macrophages play a significant role in the clearance of inhaled biotherapeutics [88, 89], including Abs [90], in the distal part of the lungs. As mentioned above, Ab exhibited a low absorption rate in the alveolar space, making them more vulnerable to macrophage uptake and subsequent degradation [91]. This situation is complicated for particles that are hard to dissolve which are cleared by alveolar macrophages phagocytosis [64].

3.2.3 Proteolytic Microenvironment

Proteases are catalytic enzymes that are critically involved in the normal function of the healthy lung. Several proteolytic enzymes have been found expressed in the lung although their absolute concentrations and activity are still debating [92–95]. Due to their pleiotropic functions and the irrevocability of their mode of action, protease activity must be tightly regulated notably by endogenous anti-proteases inhibitors. The neutrality of protease/anti-protease balance is a marker of the healthy lung. Small peptides [96] as well as large proteins, like Abs [97, 98], are sensible to extracellular proteases. Almost all lung diseases are associated with an increased expression and activity of lung proteases and a dysregulated protease/anti-protease balance [99, 100]. Acute respiratory distress syndrome (ARDS), COPD and CF are associated with increased levels of neutrophils elastase and proteinase 3 [101–105]. As exemplified by several reports demonstrating that proteases, present in the lungs, affect the integrity of Ab [106–112], it seems important to consider the sensitivity of Abs to the lung proteolytic environment during inhaled Ab drug development, as lung proteases may modify inhaled biologic stability and PK [113].

3.2.4 Endogenous Catabolism

For cell surface target, the Ab-receptor complex often eliminates like the target and depends on the Ab dose relative to the antigen expression level through TMDD. For example, at low Ab dose, the clearance of the IgG may be rapid, due to rapid internalization and elimination of the IgG-receptor complex, while at increasing dose, the receptor is saturated, and the Ab follows a more typical kinetic behaviour of an IgG. Consequently, increased expression or suppression of the soluble target or receptor may affect the duration of the Ab effect.

Inhaled Abs may undergo receptor-mediated endocytosis through the binding of the Fc domain to FcγRs, which are expressed at the surface of many immune cells. This will trigger internalization of the Ab-FcγR complex and its intracellular catabolism. However, this process is marginally involved in the overall elimination of Abs as demonstrated by experiments using FcγR knockout animals in which FcγR-mediated Ab elimination plays a limited role [114]. Besides this specific phenomenon, unbound Abs may be uptake in airway cells by pinocytosis. This non-specific and non-saturable process will promote endocytosis of small droplets of extracellular fluid containing dissolved inhaled material subsequently triggering lysosomal degradation [115]. However, the salvage pathway, provided by FcRn, will protect Abs from intracellular catabolism and promote recycling into the bloodstream of the airways after release at neutral pH into the circulation [116, 117]. The efficiency of the FcRn-mediated recycling pathway was estimated to concern ~ two-thirds of the Ab uptake in the endosome [118]. It was further confirmed in experiments using FcRn knockout animals in which Ab clearance was 10-fold accelerated [119]. However, FcRn pathway displayed saturating recycling capacity, especially in the presence of a high exogenous concentration of Abs.

4 Development of Inhaled Ab

The route of administration of a therapeutic agent has a critical impact on its efficacy. Most Ab is usually administered intravenously or through a systemic route. As exemplified in the previous sections, the systemic routes displayed a main drawback that is the limited absorption from the blood circulation to the airways. As most of the pathophysiological processes associated with respiratory diseases occur in the airways, it appears reasonable to consider the administration of Abs by inhalation. In the next section, we will provide an overview of the major development of inhaled antibodies (either as liquid aerosols or dried powders), which have been evaluated for the treatment of respiratory diseases, providing insights into their pharmacodynamics profiles.

4.1 Inhaled Abs Used for the Treatment of Respiratory Infections

Because most respiratory infections start at the mucosal surface of the upper respiratory tract, local delivery of Abs will provide frontline sterilizing passive immunity, preventing pathogen growth and dissemination. It is noteworthy that most inhaled anti-infectious Abs have been developed as a post-exposure treatment and would be most likely the most efficient if they are delivered in a short interval after pathogen exposure to prevent the occurrence of severe disease.

4.1.1 ALX-0171 and Anti-RSV Ab

RSV is a leading cause of lower respiratory tract infections in children and the elderly, with a disease burden equivalent to influenza infections [120]. While most RSV infections resolve on their own, 15–40% of immunocompromised host, especially preterm infants, and elderly, develop a more serious airway infection, which may eventually lead to bronchiolitis or even pneumonia. There is no specific treatment for RSV infection, neither vaccine nor effective antiviral drugs. The only approved drug is palivizumab, an intravenously injected humanized IgG1, targeting the RSV F protein. This neutralizing Ab prevents fusion between the viral and the host cell membrane. Despite therapeutic efficacy demonstrated in adults and prophylactic potency in children [121, 122], palivizumab has been criticized since its approval. Its cost-effectiveness has been questioned, and a consensus has emerged regarding its limited clinical benefit with no significant effect on mortality [123]. To circumvent these issues, Ablynx has developed an inhaled anti-RSV trivalent domain Ab (dAb, Nanobody®), derived from heavy chain-only abs from Camelidae. ALX-0171 is a 42-kDa Nanobody® partially targeting the same RSV epitope as palivizumab, inhibiting the release of the virus from the apical surface of bronchial epithelial cell cultures [124]. Nebulized ALX-0171 was well-tolerated and significantly reduced nasal and lung viral loads and lung lesion, to a greater extent as compared to palivizumab, in cotton rats and neonatal lambs models; the latter displaying anatomical and physiological similarities to human infants [125, 126]. These encouraging results drove the initiation of ALX-0171 clinical evaluation. The safety and tolerability of inhaled ALX-0171 was established in a first-in-human phase I/IIa clinical trial, over 60 adults (NCT01483911) and 48 infants (1–24 months old; NCT02309320) with no treatment-related serious adverse events reported. A promising reduction of global severity score was also observed in treated infants [127] and promoted the initiation of phase II studies. A significant dose-dependent reduction of viral load was observed in hospitalized infants and young children with RSV infection but without improvement of clinical outcome, including adequate oxygen saturation and oral feeding (NCT02979431) [24]. This led to the termination of the ALX-0171 program. It is noteworthy that the failure of ALX-0171 may not be associated with the route of administration but more conceptual issues including the antigen target and the population selected. In advanced RSV infections, the host dysregulated inflammatory response may drive forward a severe disease and the efficient neutralization of the virus (by the Ab) is unable to interrupt the host immune trajectory.

4.1.2 Anti-influenza Abs

Influenza infections remains a significant threat for worldwide public health with significant morbidity and mortality every year [128] and with a serious potential for devastating pandemic [129]. With the limited efficacy of vaccines, due to antigenic drift, and the limited efficacy of antivirals, due to virus resistance, there is an urgent need for the development of novel broad coverage anti-influenza therapeutics. Consequently, several Abs have been isolated and evaluated in animal or clinical studies, essentially using systemic routes of delivery. However, in most of these studies, the high amount of Ab required to protect against influenza infection is not compatible with affordable manufacturing process and healthcare system operating. Therefore, local administration, allowing the reduction of the administered dose, appears as a clinically relevant approach.

Comparative mouse studies revealed that inhalation of broadly neutralizing anti-influenza Abs (bNabs) gave a 10- to 50-fold better reduction of morbidity and mortality protection than systemic deliveries. This was associated with an improved control of the lung viral dose and inflammatory response [130, 131]. Interestingly, local administration of bNabs conferred heterosubtypic protection against divergent influenza virus subtypes [130]. Other formats, including IgY – from the yolk of chicken eggs – or Nanobodies® administered through the intranasal route, provided similar protection [132, 133].

4.1.3 Anti-SARS-CoV-2 Abs

The COVID-19 global pandemic encourages the development of therapeutics aiming at disrupting the cellular entry of the SARS-CoV-2 virus into host cells. Most of them were developed to target the receptor-binding domain (RBD) of SARS-CoV-2 spike glycoprotein, blocking the virus binding to the angiotensin-converting enzyme 2 (ACE2) receptor on the surface of host cell. RBD was used predominantly as the target in clinical-stage vaccines, which have globally demonstrated up to 95% during phase III clinical trials. However, the emergence of SARS-CoV-2 variants, and the recurrence of non-vaccinated subpopulations have raised concern about the effectiveness of the current vaccines and highlighted the necessity of alternative therapeutic strategies. Among them, Ab targeting the receptor-binding domain (RBD) may prevent viral entry, limiting its spreading throughout the body [134, 135]. For instance, at least six Abs or cocktail of Abs have been approved or received emergency use approval for the treatment of early stage vulnerable COVID-19 patients with systemic administration. The portal of entry and site of primary replication for SARS-CoV-2 is the upper respiratory tract before reaching the lungs, which are the main target organ for pathogenesis or to other individuals. Consequently, the local delivery of Abs in the airways by reducing virus dissemination and transmission could offers a tremendous opportunity to benefit to infected patients and public health.

Among them, regdanvimab (CT-P59), a recombinant human IgG1, was isolated from a screening of an antibody library constructed from peripheral blood mononuclear cells of a convalescent patient [136]. It was first approved in September 2021 and administered as a single intravenous infusion before being evaluated in combination with another neutralizing RBD-targeting Ab (CT-P63) as a nebulized treatment (NCT05224856). Interestingly, the nebulized formulation used the muco-trapping ab platform developed by Inhalon Biopharma, which enhances Ab binding to respiratory mucus, preventing the local spread of the infection by efficiently eliminating the virus through muco-ciliary clearance [137]. Using the same isolation strategy, the 1212C2 human Ab was developed and provided prophylactic and therapeutic protection when delivered parenterally in animal models. However, as an inhaled liquid aerosol, using a commercially available nebulizer, 1212C2 demonstrated a complete eradication of viral load in the nose and lungs of infected hamsters. This protection was associated with a significant dose sparing as compared to parenterally administration [138]. These results were the basis of the development of inhaled cocktail of ab targeting the spike protein and administered by inhalation [139], which was planned for clinical evaluation at the end of 2022 [140]. IBIO123 is an inhaled cocktail of three Abs binding to overlapping epitopes of the spike protein with a substantial neutralizing activity against SARS-CoV-2 variants. It is under dose-escalating phase I/II study evaluation (NCT05303376/NCT05298813). Apart from the classical full-length IgG, other Ab's format has been considered. IGM-6268, is an engineered pentameric IgM showing promising results for combatting SARS-CoV-2 and variants as intranasal therapeutics. IgM is the first line of defence against infection and has been shown to effectively neutralize hepatitis B virus (HBV), human immunodeficiency virus (HIV) or influenza viruses [141]. IGM-6268 expressed the variable regions from a potent IgG grafted on an IgM scaffold, generating an IgM with 10 binding sites of high specificity, affinity, and avidity against the spike protein of SARS-CoV-2. It is under dose-escalating phase I study evaluation (NCT05160402/NCT05184218). Nanobodies®, including Nb11-59 and PiN-21 have also been developed to target the RBD domain of the spike protein and have shown neutralizing activity even after nebulization with interesting dose minimization [142, 143]. The development of inhaled Ab treatments for SARS-CoV-2 is underway; preclinical and early clinical studies have confirmed that this approach is of particular interest for post-exposure treatment in at-risk patients to avoid severe disease and outpatient therapy [144, 145]. Finally, we hypothesize that inhaled anti-SARS-CoV-2 Ab, which have been shown to drastically reduce viral burden [138] may limit the emergence of variants, as it was associated with incomplete viral clearance after intravenous Ab treatment [146–148].

4.1.4 Anti-*Pseudomonas aeruginosa* Abs

P. aeruginosa is an opportunistic bacterium causing severe acute and persistent infections in immunocompromised individuals. Due to its highly versatile genome, this pathogen is intrinsically resistant to numerous antibiotics and has consequently been listed by the World Health Organization (WHO) as a priority pathogen. Although numerous *P. aeruginosa* antigens have been envisioned for immunotherapy or vaccination (anti-LPS, anti-flagellin, anti-pili) some of them are reaching clinical trial evaluation – their efficacy remains controversial and limited [149], which exemplifies the necessity of optimizing Abs administration. In a murine model of acute lung infection, a comparative study of the efficacy of a murine Ab recognizing the type 3 secretion system – a virulence factor expressed by *P. aeruginosa* – demonstrated a better protection associated with inhalation as compared to systemic delivery. This was associated with a better control of the lung bacterial load as well as lung inflammation [150]. In a mechanically ventilated piglets model resembling ventilator-associated pneumonia induced by *P. aeruginosa*, the nebulization of anti-*P. aeruginosa* IgY induced a transient reduction of bacterial growth associated with decreased body temperature, cardiac index and static compliance [151].

4.2 Inhaled Abs Used for the Treatment of Inflammatory Respiratory Diseases

4.2.1 Inhaled Abs for the Treatment of Asthma

Asthma is a chronic respiratory disease affecting million people worldwide and is defined as complex inflammatory syndrome encompasses heterogeneous clinical situations. Allergic asthma is the most prevalent phenotype associated with type 2 or type 17 inflammation culminating in the production of IgE by B lymphocytes and the subsequent pathologic activation of basophils and mast cells. Omalizumab (Xolair), a recombinant humanized monoclonal anti-IgE antibody that blocks the interaction of IgE with its receptors, was the first anti-IgE Ab to provide clinical success after systemic administration. Interestingly, one study evaluated the efficacy of omalizumab administered via nebulization in patients with mild allergic asthma without revealing any positive outcome on methacholine-induced bronchoconstriction nor remarkable changes in serum IgE [152]. This failure might be attributable to the low systemic concentration of omalizumab after inhalation, which cannot counteract the high-serum pathogenic IgE.

Additional proinflammatory mediators have also been considered as targets for asthma immunotherapy. Thymic stromal lymphopoietin (TSLP), an epithelial-derived cytokine produced in response to proinflammatory stimuli was shown to play an important role in allergic asthma [153]. CJS-117 is a neutralizing IgG2λ Fab fragment directed against human TSLP formulated as PulmoSol® engineered powder to be delivered via a DPI to adults with mild atopic asthma. The results of the phase I study (NCT03138811) showed that inhaled anti-TSLP was well-tolerated and associated with a reduction of both early and late asthmatic responses as compared to the placebo control group. In addition, investigators also observed a significant decrease in fractional exhaled nitric oxide (FeNO) levels throughout the study with no serious adverse effects occurring [154]. Moreover, in a phase IIa study, CSJ-117 was able to reduce allergen-induced bronchoconstriction in adult patients with mild asthma (NCT04410523/ NCT04946318).

Th2 inflammation associated with asthma is characterized by elevated levels of IL5 and IL13. The efficacy of nebulized humanized IgG1 anti-IL13 Fab fragment (CDP7766) were investigated in experimental models of allergic asthma and revealed good tolerance as well as significant suppression of airway inflammation in both mouse and cynomolgus macaque [155, 156]. Interestingly, a comparative study showed that systemic administration of anti-IL13 Fab was not protective, probably due to a short half-life [155]. Based on these results, VR942, a dry powder formulation containing CDP7766 was investigated in a phase I study (NCT02473939). Inhaled VR942 was well-tolerated with no serious adverse effect or immunogenicity as compared to placebo. In addition, preliminary evidence showed a rapid and durable inhibition of FeNO [57]. These data established inhaled Ab as a potential future therapy for asthma that is an alternative to parenteral administration.

4.2.2 Inhaled Abs for the Treatment of Acute Lung Injury

Acute lung injury (ALI) and ARDS are acute inflammatory lung diseases resulting from various processes involving directly or indirectly the airways. These diseases defined by a myriad of clinical criteria – including notably pulmonary vascular permeability, loss of aerated tissues leading to profound hypoxemia – have a high incidence and remain a significant source of morbidity and mortality in intensive care unit patients [157]. The present therapeutic approaches for ALI/ARDS include supportive care, ventilator support and corticosteroid therapy. In this context, the use of anti-inflammatory Abs to dampen excessive harmful inflammation appeared to be an attractive approach. However, despite encouraging preclinical evidence, systemic targeting of proinflammatory cytokines, including TNF-α and IL-1β, did not improve the outcome of at-risk or diagnosed patients [158]. More recently, local delivery of potential anti-inflammatory Abs via aerosol has been evaluated to optimize their effects. GSK1995057 is an inhibiting anti-TNFR1 domain antibody (dAb) developed for the prophylaxis and treatment of ALI. dAb is the smallest functional antigen binding unit derived from Ab; it comprises the variable regions of the heavy and light chains. This format was chosen to limit the tendency of full-length Ab to cross-link surface receptor, thereby activating rather than inhibiting signalling. Preclinical evaluation of inhaled GSK1995057 dAb in mouse and cynomolgus monkey ALI models showed that it significantly reduced airway inflammation as compared to full-length anti-TNFR1 Ab [159]. Phase I clinical evaluation confirmed a positive reduction of airway inflammation after LPS challenge; but unexpectedly, it revealed the pre-existence of naturally occurring anti-GSK1995057 autoantibodies in the serum of approximately 50% of patients after inhalation (NCT01587807) or systemic (NCT01476046) administration which may impact the safety and clinical pharmacology of GSK1995057 [160, 161]. A dAb derivate, GSK2862277, was developed with reduced binding to autoantibodies and evaluated in transthoracic oesophagectomy patients at-risk of developing ARDS (NCT02221037). Inhaled GSK2862277 was well-tolerated with but did not achieve a significant lowering of postoperative alveolar capillary leak despite optimal lung exposure and reduced biomarkers of lung permeability and inflammation [162]. The therapeutic potential of inhaled anti-TNFR1 dAb in ALI requires further investigations.

5 Conclusion

The pulmonary route remains rare for delivering protein therapeutics, with only few examples of approved inhaled biologics. Despite numerous promising preclinical evidences and the thriving focus of research improving our knowledge, there is no inhaled Ab product approved yet. Several explanations may be raised, as highlighted in this book chapter. Selecting appropriately the target antigen operating within the lungs and the population that may benefit from inhaled Ab is pivotal for clinical success and to pave the way for further inhaled Ab developments. Ensuring the stability of Ab during aerosolization and/or drying is mandatory to deliver a safe and efficient product into the lungs. Understanding better the pharmacological properties of the different Ab-based therapeutics after they deposit into the lungs would be valuable in selecting the most relevant Ab format to be used for a specific medical application. In addition, it is noteworthy that designing a toxicology study for inhaled Abs is not straightforward [34, 163]. There are no specific guidance or guidelines on conducting a toxicology study with inhaled Abs. Here are some examples of the issues to be considered: (1) species selection as no animal models reproduce the respiratory parameter, the lung anatomy and physiology/immunology of human ones, (2) defining the dose to deliver and how estimating the pulmonary deposited dose, (3) determining the method for aerosol generation taking into account the species, Ab instability, (4) dosing frequency as lung half-life may be different from Ab systemic half-life, (5) determining the methods/read-outs to characterize responses to inhaled Abs [65]. Moreover, there is no consensus on the interpretation of the toxicology results, making it difficult to reach an agreement with regulatory agencies on a specific inhaled Ab risk-benefit profile, and thereby, to progress into First-in-Human clinical trials. Despite the challenges associated to the pulmonary route, several inhaled Ab reached clinical trials, and we are waiting for the first clinical success that will pave the way for future inhaled Ab developments.

Acknowledgements The work associated with this review was supported by grants provided by the French National Research Agency (ANR) as part of the "Investissements d'Avenir" program (LabEx MAbImprove, ANR-10-LABX-53-01), Region Centre-Val-de-Loire (Novantinh Program), Vaincre-la-Mucoviscodose (RF20210502871) and the European Defence Fund (CounterAct program). The figures were created with Biorender.com

Permission The authors confirm having obtained permission for any material within the manuscript.

Conflict of Interest Statement TS has nothing to declare. NHV is co-founder and scientific expert for Cynbiose Respiratory. In the past 2 years, she received consultancy fees from Novartis and research support from CSL Behring, Aptar Pharma and Aerogen Ltd.

References

1. Foundation WL. Acute respiratory infections atlas. New York: World Lung Foundation; 2010.
2. Secher T, Guilleminault L, Reckamp K, Amanam I, Plantier L, Heuze-Vourc'h N. Therapeutic antibodies: a new era in the treatment of respiratory diseases? Pharmacol Ther. 2018;189:149–72.
3. Mayor A, Chesnay A, Desoubeaux G, Ternant D, Heuze-Vourc'h N, Secher T. Therapeutic antibodies for the treatment of respiratory tract infections-current overview and perspectives. Vaccines (Basel). 2021;9(2):151.
4. Lyly A, Laulajainen-Hongisto A, Gevaert P, Kauppi P, Toppila-Salmi S. Monoclonal antibodies and airway diseases. Int J Mol Sci. 2020;21(24):9477.
5. Siafakas N. Monoclonal antibodies for chronic obstructive pulmonary disease. Pulmonology. 2020;26(2):61–2.
6. Vidarsson G, Dekkers G, Rispens T. IgG subclasses and allotypes: from structure to effector functions. Front Immunol. 2014;5:520.
7. Stewart R, Hammond SA, Oberst M, Wilkinson RW. The role of Fc gamma receptors in the activity of immunomodulatory antibodies for cancer. J Immunother Cancer. 2014;2(1):29.
8. Hwang WY, Foote J. Immunogenicity of engineered antibodies. Methods. 2005;36(1):3–10.
9. Harding FA, Stickler MM, Razo J, DuBridge RB. The immunogenicity of humanized and fully human antibodies: residual immunogenicity resides in the CDR regions. MAbs. 2010;2(3):256–65.
10. Bates A, Power CA. David vs. Goliath: the structure, function, and clinical prospects of antibody fragments. Antibodies (Basel). 2019;8(2):28.
11. Kholodenko RV, Kalinovsky DV, Doronin II, Ponomarev ED, Kholodenko IV. Antibody fragments as potential biopharmaceuticals for cancer therapy: success and limitations. Curr Med Chem. 2019;26(3):396–426.
12. Gebauer M, Skerra A. Engineering of binding functions into proteins. Curr Opin Biotechnol. 2019;60:230–41.
13. Simeon R, Chen Z. In vitro-engineered non-antibody protein therapeutics. Protein Cell. 2018;9(1):3–14.
14. Sekhon B, Saluja V. Biosimilars: an overview. Biosimilars. 2011;1:11.
15. Le Basle Y, Chennell P, Tokhadze N, Astier A, Sautou V. Physicochemical stability of monoclonal antibodies: a review. J Pharm Sci. 2020;109(1):169–90.
16. Burkitt W, Domann P, O'Connor G. Conformational changes in oxidatively stressed monoclonal antibodies studied by hydrogen exchange mass spectrometry. Protein Sci. 2010;19(4):826–35.
17. Narhi LO, Schmit J, Bechtold-Peters K, Sharma D. Classification of protein aggregates. J Pharm Sci. 2012;101(2):493–8.
18. Philo JS. A critical review of methods for size characterization of non-particulate protein aggregates. Curr Pharm Biotechnol. 2009;10(4):359–72.
19. Zolls S, Tantipolphan R, Wiggenhorn M, Winter G, Jiskoot W, Friess W, et al. Particles in therapeutic protein formulations, part 1: overview of analytical methods. J Pharm Sci. 2012;101(3):914–35.
20. Bodier-Montagutelli E, Respaud R, Perret G, Baptista L, Duquenne P, Heuze-Vourc'h N, et al. Protein stability during nebulization: mind the collection step! Eur J Pharm Biopharm. 2020;152:23–34.
21. Europe Co. Particulate contamination: sub-visible particles – PDG revision of general chapter released for public consultation. In: Healthcare EDftQoM, editor. ; 2021.
22. Van Heeke G, Allosery K, De Brabandere V, De Smedt T, Detalle L, de Fougerolles A. Nanobodies(R) as inhaled biotherapeutics for lung diseases. Pharmacol Ther. 2017;169:47–56.
23. Matthews AA, Ee PLR, Ge R. Developing inhaled protein therapeutics for lung diseases. Mol Biomed. 2020;1(1):11.

24. Cunningham S, Piedra PA, Martinon-Torres F, Szymanski H, Brackeva B, Dombrecht E, et al. Nebulised ALX-0171 for respiratory syncytial virus lower respiratory tract infection in hospitalised children: a double-blind, randomised, placebo-controlled, phase 2b trial. Lancet Respir Med. 2021;9(1):21–32.
25. Hertel S, Pohl T, Friess W, Winter G. That's cool! nebulization of thermolabile proteins with a cooled vibrating mesh nebulizer. Eur J Pharm Biopharm. 2014;87(2):357–65.
26. Mayor A, Thibert B, Huille S, Respaud R, Audat H, Heuze-Vourc'h N. Inhaled antibodies: formulations require specific development to overcome instability due to nebulization. Drug Deliv Transl Res. 2021;11(4):1625–33.
27. Secher T, Bodier-Montagutelli E, Parent C, Bouvart L, Cortes M, Ferreira M, et al. Aggregates associated with instability of antibodies during aerosolization induce adverse immunological effects. Pharmaceutics. 2022;14(3):671.
28. Depreter F, Pilcer G, Amighi K. Inhaled proteins: challenges and perspectives. Int J Pharm. 2013;447(1–2):251–80.
29. Pilcer G, Amighi K. Formulation strategy and use of excipients in pulmonary drug delivery. Int J Pharm. 2010;392(1–2):1–19.
30. Hickey AJ, Stewart IE. Inhaled antibodies: quality and performance considerations. Hum Vaccin Immunother. 2022;18(2):1940650.
31. Respaud R, Marchand D, Pelat T, Tchou-Wong KM, Roy CJ, Parent C, et al. Development of a drug delivery system for efficient alveolar delivery of a neutralizing monoclonal antibody to treat pulmonary intoxication to ricin. J Control Release. 2016;234:21–32.
32. Respaud R, Marchand D, Parent C, Pelat T, Thullier P, Tournamille JF, et al. Effect of formulation on the stability and aerosol performance of a nebulized antibody. MAbs. 2014;6(5):1347–55.
33. Hall AP, Tepper JS, Boyle MH, Cary MG, Flandre TG, Piaia A, et al. BSTP review of 12 case studies discussing the challenges, pathology, immunogenicity, and mechanisms of inhaled biologics. Toxicol Pathol. 2021;49(2):235–60.
34. Wolfreys A, Kilgour J, Allen AD, Dudal S, Freke M, Jones D, et al. Review of the technical, toxicological, and PKPD considerations for conducting inhalation toxicity studies on biologic pharmaceuticals-the outcome of a cross-industry working group survey. Toxicol Pathol. 2021;49(2):261–85.
35. Jarvi NL, Balu-Iyer SV. Immunogenicity challenges associated with subcutaneous delivery of therapeutic proteins. BioDrugs. 2021;35(2):125–46.
36. Kuriakose A, Chirmule N, Nair P. Immunogenicity of biotherapeutics: causes and association with posttranslational modifications. J Immunol Res. 2016;2016:1298473.
37. Ponce R, Abad L, Amaravadi L, Gelzleichter T, Gore E, Green J, et al. Immunogenicity of biologically-derived therapeutics: assessment and interpretation of nonclinical safety studies. Regul Toxicol Pharmacol. 2009;54(2):164–82.
38. Kamath AV. Translational pharmacokinetics and pharmacodynamics of monoclonal antibodies. Drug Discov Today Technol. 2016;21-22:75–83.
39. Colombat M, Mal H, Groussard O, Capron F, Thabut G, Jebrak G, et al. Pulmonary vascular lesions in end-stage idiopathic pulmonary fibrosis: histopathologic study on lung explant specimens and correlations with pulmonary hemodynamics. Hum Pathol. 2007;38(1):60–5.
40. Hanumegowda C, Farkas L, Kolb M. Angiogenesis in pulmonary fibrosis: too much or not enough? Chest. 2012;142(1):200–7.
41. Durand RE, Aquino-Parsons C. Clinical relevance of intermittent tumour blood flow. Acta Oncol. 2001;40(8):929–36.
42. Respaud R, Vecellio L, Diot P, Heuze-Vourc'h N. Nebulization as a delivery method for mAbs in respiratory diseases. Expert Opin Drug Deliv. 2015;12(6):1027–39.
43. Hart TK, Cook RM, Zia-Amirhosseini P, Minthorn E, Sellers TS, Maleeff BE, et al. Preclinical efficacy and safety of mepolizumab (SB-240563), a humanized monoclonal antibody to IL-5, in cynomolgus monkeys. J Allergy Clin Immunol. 2001;108(2):250–7.

44. Dall'Acqua WF, Kiener PA, Wu H. Properties of human IgG1s engineered for enhanced binding to the neonatal Fc receptor (FcRn). J Biol Chem. 2006;281(33):23514–24.
45. Han A, Czajkowski L, Rosas LA, Cervantes-Medina A, Xiao Y, Gouzoulis M, et al. Safety and efficacy of CR6261 in an Influenza A H1N1 healthy human challenge model. Clin Infect Dis. 2021;73(11):e4260–8.
46. Herve V, Rabbe N, Guilleminault L, Paul F, Schlick L, Azzopardi N, et al. VEGF neutralizing aerosol therapy in primary pulmonary adenocarcinoma with K-ras activating-mutations. MAbs. 2014;6(6):1638–48.
47. Guilleminault L, Azzopardi N, Arnoult C, Sobilo J, Herve V, Montharu J, et al. Fate of inhaled monoclonal antibodies after the deposition of aerosolized particles in the respiratory system. J Control Release. 2014;196:344–54.
48. Ryman JT, Meibohm B. Pharmacokinetics of monoclonal antibodies. CPT Pharmacometrics Syst Pharmacol. 2017;6(9):576–88.
49. Folkesson HG, Westrom BR, Karlsson BW. Permeability of the respiratory tract to different-sized macromolecules after intratracheal instillation in young and adult rats. Acta Physiol Scand. 1990;139(2):347–54.
50. Conhaim RL, Watson KE, Lai-Fook SJ, Harms BA. Transport properties of alveolar epithelium measured by molecular hetastarch absorption in isolated rat lungs. J Appl Physiol (1985). 2001;91(4):1730–40.
51. Holter JF, Weiland JE, Pacht ER, Gadek JE, Davis WB. Protein permeability in the adult respiratory distress syndrome. Loss of size selectivity of the alveolar epithelium. J Clin Invest. 1986;78(6):1513–22.
52. Patton JS, Trinchero P, Platz RM. Bioavailability of pulmonary delivered peptides and proteins: α-interferon, calcitonins and parathyroid hormones. J Control Release. 1994;28(1):79–85.
53. Patton JS, Fishburn CS, Weers JG. The lungs as a portal of entry for systemic drug delivery. Proc Am Thorac Soc. 2004;1(4):338–44.
54. Patton JS, Byron PR. Inhaling medicines: delivering drugs to the body through the lungs. Nat Rev Drug Discov. 2007;6(1):67–74.
55. Labiris NR, Dolovich MB. Pulmonary drug delivery. Part I: physiological factors affecting therapeutic effectiveness of aerosolized medications. Br J Clin Pharmacol. 2003;56(6):588–99.
56. Maillet A, Guilleminault L, Lemarie E, Lerondel S, Azzopardi N, Montharu J, et al. The airways, a novel route for delivering monoclonal antibodies to treat lung tumors. Pharm Res. 2011;28(9):2147–56.
57. Burgess G, Boyce M, Jones M, Larsson L, Main MJ, Morgan F, et al. Randomized study of the safety and pharmacodynamics of inhaled interleukin-13 monoclonal antibody fragment VR942. EBioMedicine. 2018;35:67–75.
58. Patil HP, Freches D, Karmani L, Duncan GA, Ucakar B, Suk JS, et al. Fate of PEGylated antibody fragments following delivery to the lungs: influence of delivery site, PEG size and lung inflammation. J Control Release. 2018;272:62–71.
59. Niven RW. Delivery of biotherapeutics by inhalation aerosol. Crit Rev Ther Drug Carrier Syst. 1995;12(2–3):151–231.
60. Wolff RK. Safety of inhaled proteins for therapeutic use. J Aerosol Med. 1998;11(4):197–219.
61. Kim KJ, Malik AB. Protein transport across the lung epithelial barrier. Am J Physiol Lung Cell Mol Physiol. 2003;284(2):L247–59.
62. Groneberg DA, Fischer A, Chung KF, Daniel H. Molecular mechanisms of pulmonary peptidomimetic drug and peptide transport. Am J Respir Cell Mol Biol. 2004;30(3):251–60.
63. Bitonti AJ, Dumont JA. Pulmonary administration of therapeutic proteins using an immunoglobulin transport pathway. Adv Drug Deliv Rev. 2006;58(9–10):1106–18.
64. Dickinson BL, Badizadegan K, Wu Z, Ahouse JC, Zhu X, Simister NE, et al. Bidirectional FcRn-dependent IgG transport in a polarized human intestinal epithelial cell line. J Clin Invest. 1999;104(7):903–11.

65. Guillon A, Pardessus J, Lhommet P, Parent C, Respaud R, Marchand D, et al. Exploring the fate of inhaled monoclonal antibody in the lung parenchyma by microdialysis. MAbs. 2019;11(2):297–304.
66. Leal J, Smyth HDC, Ghosh D. Physicochemical properties of mucus and their impact on transmucosal drug delivery. Int J Pharm. 2017;532(1):555–72.
67. Lai SK, Wang YY, Cone R, Wirtz D, Hanes J. Altering mucus rheology to "solidify" human mucus at the nanoscale. PLoS One. 2009;4(1):e4294.
68. Thornton DJ, Rousseau K, McGuckin MA. Structure and function of the polymeric mucins in airways mucus. Annu Rev Physiol. 2008;70:459–86.
69. Rose MC, Voynow JA. Respiratory tract mucin genes and mucin glycoproteins in health and disease. Physiol Rev. 2006;86(1):245–78.
70. Sanders N, Rudolph C, Braeckmans K, De Smedt SC, Demeester J. Extracellular barriers in respiratory gene therapy. Adv Drug Deliv Rev. 2009;61(2):115–27.
71. Schuster BS, Suk JS, Woodworth GF, Hanes J. Nanoparticle diffusion in respiratory mucus from humans without lung disease. Biomaterials. 2013;34(13):3439–46.
72. Murgia X, Pawelzyk P, Schaefer UF, Wagner C, Willenbacher N, Lehr CM. Size-limited penetration of nanoparticles into Porcine respiratory mucus after aerosol deposition. Biomacromolecules. 2016;17(4):1536–42.
73. Olmsted SS, Padgett JL, Yudin AI, Whaley KJ, Moench TR, Cone RA. Diffusion of macromolecules and virus-like particles in human cervical mucus. Biophys J. 2001;81(4):1930–7.
74. Saltzman WM, Radomsky ML, Whaley KJ, Cone RA. Antibody diffusion in human cervical mucus. Biophys J. 1994;66(2 Pt 1):508–15.
75. Cone RA. Barrier properties of mucus. Adv Drug Deliv Rev. 2009;61(2):75–85.
76. Roy I, Vij N. Nanodelivery in airway diseases: challenges and therapeutic applications. Nanomedicine. 2010;6(2):237–44.
77. Yuan S, Hollinger M, Lachowicz-Scroggins ME, Kerr SC, Dunican EM, Daniel BM, et al. Oxidation increases mucin polymer cross-links to stiffen airway mucus gels. Sci Transl Med. 2015;7(276):276ra227.
78. Duncan GA, Jung J, Hanes J, Suk JS. The mucus barrier to inhaled gene therapy. Mol Ther. 2016;24(12):2043–53.
79. Boucher RC. Airway surface dehydration in cystic fibrosis: pathogenesis and therapy. Annu Rev Med. 2007;58:157–70.
80. Braeckmans K, Peeters L, Sanders NN, De Smedt SC, Demeester J. Three-dimensional fluorescence recovery after photobleaching with the confocal scanning laser microscope. Biophys J. 2003;85(4):2240–52.
81. Ensign LM, Tang BC, Wang YY, Tse TA, Hoen T, Cone R, et al. Mucus-penetrating nanoparticles for vaginal drug delivery protect against herpes simplex virus. Sci Transl Med. 2012;4(138):138ra179.
82. Maisel K, Reddy M, Xu Q, Chattopadhyay S, Cone R, Ensign LM, et al. Nanoparticles coated with high molecular weight PEG penetrate mucus and provide uniform vaginal and colorectal distribution in vivo. Nanomedicine (Lond). 2016;11(11):1337–43.
83. Lopez-Rodriguez E, Perez-Gil J. Structure-function relationships in pulmonary surfactant membranes: from biophysics to therapy. Biochim Biophys Acta. 2014;1838(6):1568–85.
84. Guagliardo R, Perez-Gil J, De Smedt S, Raemdonck K. Pulmonary surfactant and drug delivery: focusing on the role of surfactant proteins. J Control Release. 2018;291:116–26.
85. Mandal T, Konduru NV, Ramazani A, Molina RM, Larson RG. Effect of surface charge and hydrophobicity on phospholipid-nanoparticle Corona formation: a molecular dynamics simulation study. Colloid Interface Sci Commun. 2018;25:7–11.
86. Fernandes CA, Vanbever R. Preclinical models for pulmonary drug delivery. Expert Opin Drug Deliv. 2009;6(11):1231–45.
87. Ruge CA, Schaefer UF, Herrmann J, Kirch J, Canadas O, Echaide M, et al. The interplay of lung surfactant proteins and lipids assimilates the macrophage clearance of nanoparticles. PLoS One. 2012;7(7):e40775.

88. Ducreux J, Vanbever R. Crucial biopharmaceutical issues facing macromolecular candidates for inhalation: the role of macrophages in pulmonary protein clearance. Respir Drug Deliv Eur. 2007;2007:31–41.
89. Lombry C, Edwards DA, Preat V, Vanbever R. Alveolar macrophages are a primary barrier to pulmonary absorption of macromolecules. Am J Physiol Lung Cell Mol Physiol. 2004;286(5):L1002–8.
90. Koussoroplis SJ, Paulissen G, Tyteca D, Goldansaz H, Todoroff J, Barilly C, et al. PEGylation of antibody fragments greatly increases their local residence time following delivery to the respiratory tract. J Control Release. 2014;187:91–100.
91. Todoroff J, Vanbever R. Fate of nanomedicines in the lungs. Curr Opin Colloid Interface Sci. 2011;16(3):246–54.
92. Baginski L, Tachon G, Falson F, Patton JS, Bakowsky U, Ehrhardt C. Reverse transcription polymerase chain reaction (RT-PCR) analysis of proteolytic enzymes in cultures of human respiratory epithelial cells. J Aerosol Med Pulm Drug Deliv. 2011;24(2):89–101.
93. Woods A, Andrian T, Sharp G, Bicer EM, Vandera KA, Patel A, et al. Development of new in vitro models of lung protease activity for investigating stability of inhaled biological therapies and drug delivery systems. Eur J Pharm Biopharm. 2020;146:64–72.
94. Candiano G, Bruschi M, Pedemonte N, Musante L, Ravazzolo R, Liberatori S, et al. Proteomic analysis of the airway surface liquid: modulation by proinflammatory cytokines. Am J Physiol Lung Cell Mol Physiol. 2007;292(1):L185–98.
95. Nagae A, Abe M, Becker RP, Deddish PA, Skidgel RA, Erdos EG. High concentration of carboxypeptidase M in lungs: presence of the enzyme in alveolar type I cells. Am J Respir Cell Mol Biol. 1993;9(2):221–9.
96. Yang X, Ma JK, Malanga CJ, Rojanasakul Y. Characterization of proteolytic activities of pulmonary alveolar epithelium. Int J Pharm. 2000;195(1–2):93–101.
97. von Pawel-Rammingen U, Johansson BP, Bjorck L. IdeS, a novel streptococcal cysteine proteinase with unique specificity for immunoglobulin G. EMBO J. 2002;21(7):1607–15.
98. Brezski RJ, Jordan RE. Cleavage of IgGs by proteases associated with invasive diseases: an evasion tactic against host immunity? MAbs. 2010;2(3):212–20.
99. Youn YS, Kwon MJ, Na DH, Chae SY, Lee S, Lee KC. Improved intrapulmonary delivery of site-specific PEGylated salmon calcitonin: optimization by PEG size selection. J Control Release. 2008;125(1):68–75.
100. Al Ojaimi Y, Blin T, Lamamy J, Gracia M, Pitiot A, Denevault-Sabourin C, et al. Therapeutic antibodies - natural and pathological barriers and strategies to overcome them. Pharmacol Ther. 2021;233:108022.
101. Hashimoto S, Okayama Y, Shime N, Kimura A, Funakoshi Y, Kawabata K, et al. Neutrophil elastase activity in acute lung injury and respiratory distress syndrome. Respirology. 2008;13(4):581–4.
102. Pandey KC, De S, Mishra PK. Role of proteases in chronic obstructive pulmonary disease. Front Pharmacol. 2017;8:512.
103. Thulborn SJ, Mistry V, Brightling CE, Moffitt KL, Ribeiro D, Bafadhel M. Neutrophil elastase as a biomarker for bacterial infection in COPD. Respir Res. 2019;20(1):170.
104. Witko-Sarsat V, Cramer EM, Hieblot C, Guichard J, Nusbaum P, Lopez S, et al. Presence of proteinase 3 in secretory vesicles: evidence of a novel, highly mobilizable intracellular pool distinct from azurophil granules. Blood. 1999;94(7):2487–96.
105. Oriano M, Amati F, Gramegna A, De Soyza A, Mantero M, Sibila O, et al. Protease-antiprotease imbalance in bronchiectasis. Int J Mol Sci. 2021;22(11):5996.
106. Biancheri P, Brezski RJ, Di Sabatino A, Greenplate AR, Soring KL, Corazza GR, et al. Proteolytic cleavage and loss of function of biologic agents that neutralize tumor necrosis factor in the mucosa of patients with inflammatory bowel disease. Gastroenterology. 2015;149(6):1564–1574 e1563.

107. Curciarello R, Sobande T, Jones S, Giuffrida P, Di Sabatino A, Docena GH, et al. Human neutrophil elastase proteolytic activity in ulcerative Colitis Favors the loss of function of therapeutic monoclonal antibodies. J Inflamm Res. 2020;13:233–43.
108. Deveuve Q, Lajoie L, Barrault B, Thibault G. The proteolytic cleavage of therapeutic monoclonal antibody Hinge region: more than a matter of subclass. Front Immunol. 2020;11:168.
109. Kinder M, Greenplate AR, Grugan KD, Soring KL, Heeringa KA, McCarthy SG, et al. Engineered protease-resistant antibodies with selectable cell-killing functions. J Biol Chem. 2013;288(43):30843–54.
110. Hsiao HC, Fan X, Jordan RE, Zhang N, An Z. Proteolytic single hinge cleavage of pertuzumab impairs its Fc effector function and antitumor activity in vitro and in vivo. Breast Cancer Res. 2018;20(1):43.
111. Fan X, Brezski RJ, Fa M, Deng H, Oberholtzer A, Gonzalez A, et al. A single proteolytic cleavage within the lower hinge of trastuzumab reduces immune effector function and in vivo efficacy. Breast Cancer Res. 2012;14(4):R116.
112. Zhang N, Deng H, Fan X, Gonzalez A, Zhang S, Brezski RJ, et al. Dysfunctional antibodies in the tumor microenvironment associate with impaired anticancer immunity. Clin Cancer Res. 2015;21(23):5380–90.
113. Agu RU, Ugwoke MI, Armand M, Kinget R, Verbeke N. The lung as a route for systemic delivery of therapeutic proteins and peptides. Respir Res. 2001;2(4):198–209.
114. Abuqayyas L, Balthasar JP. Application of knockout mouse models to investigate the influence of FcgammaR on the tissue distribution and elimination of 8C2, a murine IgG1 monoclonal antibody. Int J Pharm. 2012;439(1–2):8–16.
115. Hastings RH, Folkesson HG, Matthay MA. Mechanisms of alveolar protein clearance in the intact lung. Am J Physiol Lung Cell Mol Physiol. 2004;286(4):L679–89.
116. Ober RJ, Martinez C, Lai X, Zhou J, Ward ES. Exocytosis of IgG as mediated by the receptor, FcRn: an analysis at the single-molecule level. Proc Natl Acad Sci U S A. 2004;101(30):11076–81.
117. Ko SY, Pegu A, Rudicell RS, Yang ZY, Joyce MG, Chen X, et al. Enhanced neonatal Fc receptor function improves protection against primate SHIV infection. Nature. 2014;514(7524):642–5.
118. Kim J, Hayton WL, Robinson JM, Anderson CL. Kinetics of FcRn-mediated recycling of IgG and albumin in human: pathophysiology and therapeutic implications using a simplified mechanism-based model. Clin Immunol. 2007;122(2):146–55.
119. Junghans RP, Anderson CL. The protection receptor for IgG catabolism is the beta2-microglobulin-containing neonatal intestinal transport receptor. Proc Natl Acad Sci U S A. 1996;93(11):5512–6.
120. Falsey AR, Walsh EE. Respiratory syncytial virus infection in adults. Clin Microbiol Rev. 2000;13(3):371–84.
121. Feltes TF, Cabalka AK, Meissner HC, Piazza FM, Carlin DA, Top FH Jr, et al. Palivizumab prophylaxis reduces hospitalization due to respiratory syncytial virus in young children with hemodynamically significant congenital heart disease. J Pediatr. 2003;143(4):532–40.
122. Palivizumab, a humanized respiratory syncytial virus monoclonal antibody, reduces hospitalization from respiratory syncytial virus infection in high-risk infants. The IMpact-RSV Study Group. Pediatrics. 1998;102(3 Pt 1):531–7.
123. American Academy of Pediatrics Committee on Infectious D, American Academy of Pediatrics Bronchiolitis Guidelines C. Updated guidance for palivizumab prophylaxis among infants and young children at increased risk of hospitalization for respiratory syncytial virus infection. Pediatrics. 2014;134(2):e620–38.
124. Broadbent L, Parke HG, Ferguson LJ, Millar A, Shields MD, Detalle L, et al. Comparative therapeutic potential of ALX-0171 and Palivizumab against respiratory syncytial virus clinical isolate infection of well-differentiated primary pediatric bronchial epithelial cell cultures. Antimicrob Agents Chemother. 2020;64(2):e02034.

125. Detalle L, Stohr T, Palomo C, Piedra PA, Gilbert BE, Mas V, et al. Generation and characterization of ALX-0171, a potent novel therapeutic nanobody for the treatment of respiratory syncytial virus infection. Antimicrob Agents Chemother. 2016;60(1):6–13.
126. Larios Mora A, Detalle L, Gallup JM, Van Geelen A, Stohr T, Duprez L, et al. Delivery of ALX-0171 by inhalation greatly reduces respiratory syncytial virus disease in newborn lambs. MAbs. 2018;10(5):778–95.
127. Positive top line results from ALX-0171 infant RSV study. https://www.europeanpharmaceuticalreview.com/news/40828/alx-0171-infant-rsv-study/, 2016. Accessed Date 2016.
128. Influenza seasonal. https://www.who.int/health-topics/influenza-seasonal#tabtab_1., 2020, Accessed Date Accessed 2020 Accessed.
129. Fineberg HV. Pandemic preparedness and response--lessons from the H1N1 influenza of 2009. N Engl J Med. 2014;370(14):1335–42.
130. Leyva-Grado VH, Tan GS, Leon PE, Yondola M, Palese P. Direct administration in the respiratory tract improves efficacy of broadly neutralizing anti-influenza virus monoclonal antibodies. Antimicrob Agents Chemother. 2015;59(7):4162–72.
131. Vigil A, Frias-Staheli N, Carabeo T, Wittekind M. Airway delivery of anti-influenza monoclonal antibodies results in enhanced antiviral activities and enables broad-coverage combination therapies. J Virol. 2020;94(22):e00052.
132. Ibanez LI, De Filette M, Hultberg A, Verrips T, Temperton N, Weiss RA, et al. Nanobodies with in vitro neutralizing activity protect mice against H5N1 influenza virus infection. J Infect Dis. 2011;203(8):1063–72.
133. Abbas AT, El-Kafrawy SA, Sohrab SS, Azhar EIA. IgY antibodies for the immunoprophylaxis and therapy of respiratory infections. Hum Vaccin Immunother. 2019;15(1):264–75.
134. Tortorici MA, Beltramello M, Lempp FA, Pinto D, Dang HV, Rosen LE, et al. Ultrapotent human antibodies protect against SARS-CoV-2 challenge via multiple mechanisms. Science. 2020;370(6519):950–7.
135. Lv Z, Deng YQ, Ye Q, Cao L, Sun CY, Fan C, et al. Structural basis for neutralization of SARS-CoV-2 and SARS-CoV by a potent therapeutic antibody. Science. 2020;369(6510):1505–9.
136. Kim C, Ryu DK, Lee J, Kim YI, Seo JM, Kim YG, et al. A therapeutic neutralizing antibody targeting receptor binding domain of SARS-CoV-2 spike protein. Nat Commun. 2021;12(1):288.
137. McSweeney M, Stewart I, Richardson Z, Kang H, Park Y, Kim C, et al. Stable nebulization and muco-trapping properties of Regdanvimab/IN-006 supports its development as a potent, dose-saving inhaled therapy for COVID-19. bioRxiv. 2022;
138. Piepenbrink MS, Park JG, Oladunni FS, Deshpande A, Basu M, Sarkar S, et al. Therapeutic activity of an inhaled potent SARS-CoV-2 neutralizing human monoclonal antibody in hamsters. Cell Rep Med. 2021;2(3):100218.
139. Piepenbrink MS, Park J-G, Desphande A, Loos A, Ye C, Basu M, et al. Potent universal-coronavirus therapeutic activity mediated by direct respiratory administration of a Spike S2 domain-specific human neutralizing monoclonal antibody. bioRxiv. 2022;
140. Pharmaceuticals A. Aridis pharmaceuticals announces first Quarter 2022 financial results and business update, 2022.
141. Keyt BA, Baliga R, Sinclair AM, Carroll SF, Peterson MS. Structure, function, and therapeutic use of IgM antibodies. Antibodies (Basel). 2020;9(4):53.
142. Gai J, Ma L, Li G, Zhu M, Qiao P, Li X, et al. A potent neutralizing nanobody against SARS-CoV-2 with inhaled delivery potential. MedComm (2020). 2021;2(1):101–13.
143. Nambulli S, Xiang Y, Tilston-Lunel NL, Rennick LJ, Sang Z, Klimstra WB, et al. Inhalable Nanobody (PiN-21) prevents and treats SARS-CoV-2 infections in Syrian hamsters at ultra-low doses. Sci Adv. 2021;7(22):eabh0319.
144. Eedara BB, Alabsi W, Encinas-Basurto D, Polt R, Ledford JG, Mansour HM. Inhalation delivery for the treatment and prevention of COVID-19 infection. Pharmaceutics. 2021;13(7):1077.

145. McSweeney MD, Stewart I, Richardson Z, Kang H, Park Y, Kim C, et al. Stable nebulization and muco-trapping properties of regdanvimab/IN-006 support its development as a potent, dose-saving inhaled therapy for COVID-19. Bioeng Transl Med. 2022;8:e10391.
146. Jensen B, Luebke N, Feldt T, Keitel V, Brandenburger T, Kindgen-Milles D, et al. Emergence of the E484K mutation in SARS-COV-2-infected immunocompromised patients treated with bamlanivimab in Germany. Lancet Reg Health Eur. 2021;8:100164.
147. Fenaux H, Gueneau R, Chaghouri A, Henry B, Mouna L, Roque-Afonso AM, et al. Emergence of SARS-CoV-2 resistance mutations in a patient who received anti-SARS-COV2 spike protein monoclonal antibodies: a case report. BMC Infect Dis. 2021;21(1):1223.
148. Rockett R, Basile K, Maddocks S, Fong W, Agius JE, Johnson-Mackinnon J, et al. Resistance mutations in SARS-CoV-2 delta variant after Sotrovimab use. N Engl J Med. 2022;386(15):1477–9.
149. Doring G, Pier GB. Vaccines and immunotherapy against Pseudomonas aeruginosa. Vaccine. 2008;26(8):1011–24.
150. Secher T, Dalonneau E, Ferreira M, Parent C, Azzopardi N, Paintaud G, et al. In a murine model of acute lung infection, airway administration of a therapeutic antibody confers greater protection than parenteral administration. J Control Release. 2019;303:24–33.
151. Otterbeck A, Hanslin K, Lantz EL, Larsson A, Stalberg J, Lipcsey M. Inhalation of specific anti-Pseudomonas aeruginosa IgY antibodies transiently decreases P. aeruginosa colonization of the airway in mechanically ventilated piglets. Intensive Care Med Exp. 2019;7(1):21.
152. Fahy JV, Cockcroft DW, Boulet LP, Wong HH, Deschesnes F, Davis EE, et al. Effect of aerosolized anti-IgE (E25) on airway responses to inhaled allergen in asthmatic subjects. Am J Respir Crit Care Med. 1999;160(3):1023–7.
153. West EE, Kashyap M, Leonard WJ. TSLP: a key regulator of asthma pathogenesis. Drug Discov Today Dis Mech. 2012;9(3-4):10.1016/j.ddmec.2012.09.003.
154. Gauvreau G, Hohlfeld J, Boulet L-P, Cockcroft D, Davis B, Fitzgerald JM, et al. Late Breaking Abstract - efficacy of CSJ117 on allergen-induced asthmatic responses in mild atopic asthma patients. Eur Respir J. 2020;56:3690.
155. Hacha J, Tomlinson K, Maertens L, Paulissen G, Rocks N, Foidart JM, et al. Nebulized anti-IL-13 monoclonal antibody Fab' fragment reduces allergen-induced asthma. Am J Respir Cell Mol Biol. 2012;47(5):709–17.
156. Lightwood D, Tservistas M, Zehentleitner M, Sarkar K, Turner A, Bracher M, et al. Efficacy of an inhaled IL-13 antibody fragment in a model of chronic asthma. Am J Respir Crit Care Med. 2018;198(5):610–9.
157. Force ADT, Ranieri VM, Rubenfeld GD, Thompson BT, Ferguson ND, Caldwell E, et al. Acute respiratory distress syndrome: the Berlin Definition. JAMA. 2012;307(23):2526–33.
158. Standiford TJ, Ward PA. Therapeutic targeting of acute lung injury and acute respiratory distress syndrome. Transl Res. 2016;167(1):183–91.
159. Bertok S, Wilson MR, Morley PJ, de Wildt R, Bayliffe A, Takata M. Selective inhibition of intra-alveolar p55 TNF receptor attenuates ventilator-induced lung injury. Thorax. 2012;67(3):244–51.
160. Cordy JC, Morley PJ, Wright TJ, Birchler MA, Lewis AP, Emmins R, et al. Specificity of human anti-variable heavy (VH) chain autoantibodies and impact on the design and clinical testing of a VH domain antibody antagonist of tumour necrosis factor-alpha receptor 1. Clin Exp Immunol. 2015;182(2):139–48.
161. Holland MC, Wurthner JU, Morley PJ, Birchler MA, Lambert J, Albayaty M, et al. Autoantibodies to variable heavy (VH) chain Ig sequences in humans impact the safety and clinical pharmacology of a VH domain antibody antagonist of TNF-alpha receptor 1. J Clin Immunol. 2013;33(7):1192–203.

162. Ryan J, Bayliffe AI, McAuley DF, Yeung J, Thickett DR, Howells PA, et al. A nebulised antitumour necrosis factor receptor-1 domain antibody in patients at risk of postoperative lung injury: a randomised, placebo-controlled pilot study. Eur J Anaesthesiol. 2020;37(11):1014–24.
163. Secher T, Bodier-Montagutelli E, Guillon A, Heuze-Vourc'h N. Correlation and clinical relevance of animal models for inhaled pharmaceuticals and biopharmaceuticals. Adv Drug Deliv Rev. 2020;167:148–69.
164. Lonberg N. Fully human antibodies from transgenic mouse and phage display platforms. Curr Opin Immunol. 2008;20(4):450–9.

Dry Powder Formulation of Monoclonal Antibodies for Pulmonary Delivery

Kimberly B. Shepard, David Zeigler, W. Brett Caldwell, and Matthew Ferguson

Abstract Interest in monoclonal antibody therapy for lung indications has grown substantially in the last decade. Local delivery of monoclonal antibody therapies directly to lung tissue via inhalation holds potential to improve treatment outcomes by reducing dose and direct targeting of diseased tissue. This review chapter focuses on respiratory administration of monoclonal antibodies delivered by dry powder inhalers. After an introduction to the topic, requirements for a dry powder monoclonal antibody product are discussed. Manufacturing techniques used to produce these dry powders are reviewed, and an overview of formulation strategies is provided. Process scale-up, formulation-device interactions, and packaging requirements are considered. A range of relevant preclinical and clinical studies from the literature involving dry powder antibodies demonstrate progress in this field. Finally, the authors give their outlook on this promising delivery paradigm for monoclonal antibodies.

Keywords Dry powder inhaler · Spray drying · Monoclonal antibody · Pulmonary delivery

1 Introduction and Background

This review will focus on the advances in knowledge and frontiers for dry powder pulmonary delivery of monoclonal antibodies (mAbs). Some background required to support the review of this field will include tenets of pulmonary delivery,

K. B. Shepard (✉) · W. B. Caldwell · M. Ferguson
Lonza, Bend, OR, USA
e-mail: kimberly.shepard@lonza.com

D. Zeigler
Absci, Vancouver, WA, USA

J. K. W. Lam, P. C. L. Kwok (eds.), *Respiratory Delivery of Biologics, Nucleic Acids, and Vaccines*, AAPS Introductions in the Pharmaceutical Sciences 8,
https://doi.org/10.1007/978-3-031-47567-2_3

processing of pharmaceutical powders for inhalation, and general formulation knowledge for biologics.

Pulmonary delivery of aerosolized active pharmaceutical ingredients (APIs) has been explored in various forms for centuries. Despite its widespread adoption, challenges remain for pulmonary delivery in the areas of ease of device use and reliable dosing, both impacting patient compliance [1]. The development of metered dose inhalers (MDIs) in the 1950s for delivery of atomized solutions created a paradigm shift in the field since they were small, inexpensive, and could be used at home [2]. However, incorrect MDI operations resulting in improper dosing by patients led to the rise of dry powder inhalers (DPIs) as an alternative technology in the 1980s [3]. Unlike MDIs, DPIs utilize aerosolized powders and rely on patient inhalation for delivery, eliminating the complex synchronization of patient breath and device actuation associated with MDIs. An additional advantage of DPIs compared with MDIs is that they are propellant-free.

Pulmonary delivery of proteins and nebulizer delivery of mAb are discussed extensively in other chapters of this book. Nebulizers have been primarily used as the standard pulmonary delivery device for mAb therapeutics. In recent years, delivery by DPI has been studied as a way to reduce the treatment burden on the patient while enabling product storage at ambient temperatures instead of frozen solutions. This review will discuss the advantages and challenges of dry powder pulmonary delivery of mAbs.

Regulatory requirements for inhaled products

Any DPI product filing (small molecule or biologic) involves meeting various regulatory quality and efficacy requirements, which differ somewhat by region. For all DPI products, the drug product and its device are treated as a combination product. Additional regulations and/or critical quality attributes (CQAs) apply to inhaled biologic products, including future pulmonary mAb products. These guidances must be met across agencies such as the Food & Drug Administration (FDA) and European Medicines Agency (EMA) to enable worldwide sales of a product [4, 5]. The specifications generally accepted as the union of these guidances are represented in Table 1. Note that "Additional tests likely for inhaled mAb" is speculative, as no inhaled mAb products have yet achieved approved status.

In the preclinical studies referenced later in this chapter, there is a strong focus on evaluating the biologic activity, degradation, aggregation and aerodynamic particle size distribution (APSD) quality attributes. Tests that are most relevant for later-stage clinical products, such as delivered dose uniformity, microbial limits, and moisture content, will not be expanded upon in this chapter.

Table 1 Quality attributes for inhaled pharmaceutical products [5, 6]

Quality attribute	Small molecule inhaled products, US (FDA)	Small molecule inhaled products, EU (EMA)	Additional tests likely for inhaled mAb	Common methods for testing
Description	X	X		Visual
Identification	X			Fourier-transform infrared spectroscopy (FTIR), UV-vis spectroscopy, retention time (by liquid chromatography or gel electrophoresis)
Assay	X	X		UV-vis spectroscopy, retention time (liquid chromatography or gel electrophoresis)
Biological activity			X	API specific (surface plasmon resonance, ELISA or cell-based assay)
Impurities and degradation products	X	X		Liquid chromatography or gel electrophoresis
Aggregation			X	Size-exclusion chromatography (SEC) and/or dynamic light scattering
Aerodynamic particle size distribution	X	X		Next Generation Impactor, Fast-Screening Impactor, breath actuators
Delivered Dose Uniformity (DDU)	X	X		DDU apparatus USP<601>
Mean delivered dose		X		
Number of actuations per container		X		
Foreign particulate matter	X			Optical, laser diffraction
Microbial limits	X	X		
Moisture content	X	X		Karl Fischer titration
Net content weight	X			

A primary focus for any inhaled pulmonary product is the efficient delivery of the API to the target region of the airway. For inhaled powders, the main critical quality attribute is APSD. The aerodynamic diameter of a particle is equal to the diameter of a unit-density sphere whose inertial settling velocity is equivalent [6]. It is measured by cascade impaction (e.g. Next Generation Impactor (NGI), Anderson

Cascade Impactor, or Fast-Screening Impactor), and provides an assessment of how much active is delivered to the lung (assuming a 5 micron upper limit for lung delivery). Delivered dose uniformity, mean dose delivered, and number of actuations are also key measures to ensure an efficacious dose will reach patients.

Another primary consideration when evaluating an inhalation product involves the product's potential immunogenicity. One test toward this risk is the foreign particulate matter measurement. These particles can come from excipients or the manufacturing process, but with mAb products there is an added complication of stability issues resulting in protein aggregation and potential for increased immunogenicity. The challenges and considerations associated with protein aggregates in inhaled biologics were recently highlighted by Ibrahim et al. [7]. Additionally, the microbial testing includes specific microbes of interest to inhaled delivery, and these specifications are an order of magnitude below what is common for oral products as described in USP <1111>.

2 Manufacturing Process Overview

Given the strict requirements for pulmonary delivery of powders, precise control over the manufacturing process is critical to achieving the desired aerosolization properties [8, 9]. In this section, we discuss the strategies for manufacturing any inhalation dry powders with a focus on biologics DPIs and specifically mAbs.

A common method to make aerosolizable powders of small molecules is mechanical milling. While this strategy is simple and cost-effective, it can be challenging due to the generation of electrostatic charges, inconsistent morphologies, and high surface energies that promote instability [10]. For proteins, milling is not a feasible particle engineering technique: mechanical and thermal forces from milling can denature proteins and impair their activity, and require API supply in the solid state [11]. Instead, particle engineering to deliver biotherapeutics to the lung is better accomplished via drying processes that start from a liquid feedstock. This section will introduce many drying techniques while focusing on spray drying, spray-freeze drying, and thin film freezing, which were also addressed in more detail in a recent review article [12].

2.1 Spray Drying

Developed originally in the early twentieth century for the production of powdered milk products in the dairy industry, spray drying is a scalable manufacturing technique that has become a common unit operation in many industries [13]. A schematic of the spray drying process is shown in Fig. 1. In a pharmaceutical spray drying process, API and excipients are co-dissolved in a volatile solvent. The liquid feed is pumped into a drying chamber through an atomizer, where small liquid

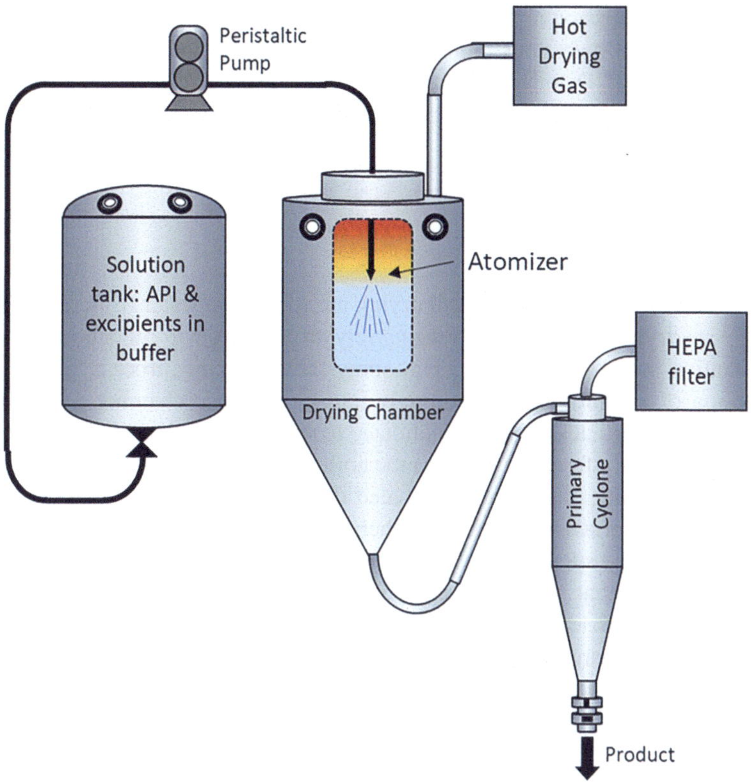

Fig. 1 Schematic of the spray drying process

droplets are formed. The droplets come into contact with heated drying gas, which rapidly removes solvent from the droplets. In this way, solid particles are formed and subsequently collected from the outlet gas stream using cyclonic separators, filter banks, or a combination thereof. Notably, spray drying has been employed in manufacturing of at least 24 commercially-approved products [14]. Specifically for particle engineering toward inhalable therapies, the use of spray drying as a particle engineering tool has been extensively reviewed by Vehring [6].

For the manufacture of dry powder inhalation dosage forms, spray drying is particularly valuable as it accomplishes both the particle engineering and formulation steps in a single unit operation. The use of spray drying to manufacture dry powders for pulmonary drug delivery has been reviewed multiple times over the past 15 years [15–17]. Spray-dried products for pulmonary delivery include: Tobi Podhaler (tobramycin, DPI), Inbrija (levadopa, DPI), Exubera (insulin, DPI), Afrezza (insulin, DPI), Cayston (aztreonam, reconstituted and nebulized), Bronchitol (mannitol, DPI), Aerovanc (vancomycin, DPI), and Inavir (laninamivir, DPI) [14, 16, 18].

Most of the discussion in this chapter will focus on spray drying as a manufacturing technique, as it represents the majority of the work in the field of inhaled biologic development. Detailed process and formulation considerations for spray drying of mAbs are discussed later in this chapter.

2.2 *Alternate Manufacturing Processes*

Rather than relying on forced, heated drying gas to remove water from the solution, spray-freeze drying and thin film freezing rely on cooled environments, similar to a lyophilization-based drying step. The spray-freeze drying process begins with atomization of a solution stream directly into a cryogenic chamber to freeze the particles. These particles are then lyophilized to remove the solvent and isolate solid particles [19]. Spray-freeze drying has the ability to create formulations with particle sizes and densities amenable to inhaled delivery. A recent review covered the application of this process to inhaled biotherapeutics in detail [20]. In an early study, Maa et al. compared spray drying and spray-freeze drying technologies for an anti-IgE mAb, finding that spray-freeze drying produced larger, low-density particles with high fine particle fraction [21]. Scale-up to clinical and commercial-scale equipment is underway for spray-freeze drying of pharmaceutical products (including an aseptic system, Lynfinity, now commercially available), but it is not yet as readily scalable or energy-efficient as spray drying at this time.

Thin film freezing (TFF) is a recently-developed platform to achieve small respirable particles of delicate molecules such as monoclonal antibodies [22]. In TFF, a liquid protein solution is applied to a cryogenic drum and rapidly frozen. Next, lyophilization is performed to remove the solvent, forming brittle matrix particles that can aerosolize upon inhalation. TFF has shown promise, particularly for particle engineering of proteins that are sensitive to air-water interfaces during atomization such as lactoferrin [23].

In a variation on the traditional spray drying process, electrostatic spray drying applies a charge to the liquid droplets as they form within the atomizer. The charge aids in driving water out of the atomized droplets, reducing the need for heated atomization gas. Thus, electrostatic atomization could help reduce water content in a powdered product while potentially reducing thermal exposure in the droplet. Mutukuri et al. employed electrostatic spray drying to manufacture inhalable dry powders of trastuzumab formulated with trehalose, phosphate buffer, and polysorbate 20 [24].

Since there are multiple processes/mechanisms to make inhalable solid particles, the most appropriate manufacturing process should be selected with the patient and product/process scalability in mind. Formulation selection and the resultant particle morphology from commercially-scalable equipment are likely the strongest driving factors for product performance. In the current state, spray drying meets these requirements and will therefore be the focus of the remainder of this chapter.

3 Formulation of Spray-Dried Inhaled MAbs

A challenge common to all dry powder manufacturing strategies is the structural denaturation of proteins during manufacturing and/or upon product storage. Many of the challenges associated with dry powder mAbs for pulmonary delivery also apply to spray-dried mAbs for other applications such as reconstitution and subsequent injection, so numerous studies on this topic are instructive and included herein. A recent review covers the role of stabilizing excipients in biomacromolecule dry powder formulations in great detail [25]. Excipients for dry powder formulation of biotherapeutics can be divided into two general categories: stabilizers and surface-active excipients. Both stabilizers and surface-active excipients have the added constraint that they must be acceptable for use in the sensitive respiratory tract. A summary of the current status of common excipients and their functions is provided in Table 2.

3.1 *Stabilizing Excipients*

As discussed in the process sections below, mAbs in dry powder inhaled formulations undergo various stresses depending on their manufacturing technique, including dehydration, shear from atomization, and air-liquid interfacial stress for spray

Table 2 Common excipients investigated for dry powder antibody formulations

Excipient	Suggested function in dry powder mAb formulation	Precedence in FDA-approved pulmonary product?
Trehalose	Stabilizer	No
Mannitol	Stabilizer	Yes
Lactose	Stabilizer	Yes
Glucose	Stabilizer	No
Sorbitol	Stabilizer	No
Cyclodextrins	Stabilizer	No
Raffinose	Stabilizer	No
L-leucine	Surface active	No
Tri-leucine	Surface active	No
Polysorbate 80	Surface active	Yes
Glycine	Stabilizer	Yes
Sodium lauryl Sulfate	Surface active	Yes
Cysteine	Stabilizer	No
Phenylalanine	Stabilizer	No
Arginine	Stabilizer	No
Polysorbate 20	Surface active	No

drying [26]. Freezing stress is encountered additionally for spray-freeze drying. On the other hand, thin film freezing eliminates the risk of shear from atomization. In these cases, excipients help protect antibodies from these processing stresses.

3.1.1 Stabilizing Sugars

Literature studies of intravenous and nebulized pulmonary formulations of mAbs focus on dissolving the active in a buffer, typically with a sugar-based excipient and a surfactant to reduce solution-state aggregation. As with intravenous (IV) formulations, sugars and sugar alcohols most often serve as stabilizing excipients for dry powder formulations, where they help in maintaining the protein's native conformation via hydrogen bonding/water replacement and reducing mobility in the solid [27].

Mannitol, lactose, sorbitol, trehalose, raffinose, glucose, and cyclodextrin are common sugar matrix excipients employed in clinical and preclinical trial formulations [25, 28–38]. Of these excipients, trehalose has emerged as the most prevalent stabilizer for dry powder inhalation delivery of biotherapeutics [39–42]. Trehalose is a non-reducing sugar (i.e., a sugar with no free ketone or aldehyde group) that is not subject to the Maillard reaction mechanism of degradation [43]. Trehalose also has a high glass transition temperature (~120 °C) under dry conditions, leading to good chemical and physical stability [25]. Materials with glass transition temperatures far above their storage temperatures will have low molecular mobility, effectively "locking in" the formulation's molecular structure, and reducing changes over time.

Sane et al. compared the processes of lyophilization and spray drying using a model monoclonal antibody (rhuMAb) with or without sucrose. For spray-dried samples, increasing the molar ratio of sucrose:rhuMAb from 55:1 to 320:1 decreased the aggregation rate fivefold during storage at 30 °C [36]. Assuming a mass of 150 kDa, the 320:1 ratio would correspond to roughly 55% antibody in the dry powder by mass. A high active loading such as this could enable a wide range of dosing strengths, up to approximately 15–18 mg active in a DPI.

Andya et al. spray-dried rhuMAbE25, an anti-IgE antibody, in formulations composed of trehalose, lactose, or mannitol from sugar:rhuMAbE25 molar ratios of 100:1 to 900:1 [44]. These formulations were then assayed for aggregation following reconstitution by size-exclusion chromatography (SEC) and the fine particle fraction (FPF, fraction of particles with mass median aerodynamic diameters <6.4 μm) by using a four-stage inertial impactor. The powders were also stored at 5, 30, and 40 °C for 1 year. All trehalose:rhuMAbE25 molar ratios above 100:1 minimized aggregation (<1% aggregate content vs. 5% without sugar), but ratios above 200:1 decreased FPF from 30–35% to 10–15%. In contrast, all lactose:rhuMAbE25 molar ratios from 100:1 to 600:1 exhibited an FPF of ~30% and had aggregation <1%. However, storage of lactose particles for 9 months at 30 °C led to irreversible glycation through Maillard reactions that were not observed during storage at

5 °C. Mannitol formulations all had higher aggregate content (2–3%) and further increasing mannitol steadily decreased FPF. All formulations decreased aggregation rate constants following storage of powders for 1 year at 30 °C. The benefit of mannitol plateaued at sugar:Ab ratios >100:1, while increasing trehalose and lactose content beyond 100:1 continued to decrease the aggregation rate constant. This observation was corroborated by powder X-ray diffraction (PXRD). After 1 year at 30 °C, trehalose at 900:1 and lactose at 600:1 showed no signs of crystallinity. In contrast, the 500:1 mannitol formulation showed a clear crystalline fingerprint. Although the 200:1 and 100:1 formulations did not exhibit crystallinity by PXRD, the extent of crystallization may have been at levels below the detection limit of the assay. These results emphasize that there may be compromises implicit in simultaneously optimizing both stability and powder properties, especially depending on the identity and/or amounts of the excipient.

Schüle et al. evaluated solutions containing 0–80% mannitol for spray drying a model IgG antibody [45]. Higher mannitol content (50–80%) led to an increase in aggregates as measured by SEC and turbidity, while 20–30% mannitol formulations were identical to 100% IgG solution. They compared aggregation in solutions comprising 0%, 30%, and 70% mannitol before and after one week at 40 °C. The 0% and 30% solutions initially had ~0.5% aggregate content while the 70% formulation started at 3%, indicating higher mannitol concentrations destabilize protein structure upon formulation. After 1 week in solution, the 30% mannitol mixture had unchanged aggregate content, while the 0% and 70% mixtures increased to 1.5% and 6%, respectively. After spray drying and powder dissolution, the powders contained 0.7–6.4% aggregates; lower mannitol content (20–40%) maximized stability during spray drying, while higher mannitol content promoted accelerated aggregation, echoing the results from the Andya et al. studies. FTIR analysis of the spray-dried powders showed a different amide I band fingerprint compared to the IgG feed solution. Interestingly, after reconstitution of the SDI powders, the solutions reverted to the original IR profile, suggesting reversible denaturation occurs during drying [45]. The reversibility and magnitude of denaturation during drying may vary according to the antibody structure and sequence. This could have important implications for clinical development as denaturation could disrupt antigen binding and lead to aggregation, which may lead to diminished efficacy and subsequent immunogenicity in vivo [7].

It is the authors' opinion, based on available evidence, that to maximize the stabilizing impact of sugar, the protein and sugar molecules must be in intimate contact. Thus, manufacturing a single-phase amorphous dispersion of sugar and protein should be prioritized. Trehalose is the foremost example of a nonreducing sugar that is often physically stable in an amorphous dispersion with a protein. Studies have also shown mannitol to be a useful stabilizing excipient, though it is prone to crystallization during spray drying under many conditions and formulation compositions. An amorphous dispersion of mannitol and protein would be preferable to maximize molecular interaction, but challenging to manufacture.

3.1.2 Stabilizing Amino Acids

Amino acids have also been assessed as DPI mAb excipients, likely serving a variety of functions dependent on the side chain functionality. In collaboration with UCB Pharma, the Massant et al. spray-dried two model IgG antibodies formulated with combinations of sucrose, trehalose, and an amino acid (13 total screened) at varying concentrations [46]. They found that stability was almost entirely dictated by the sugar; increasing sucrose or trehalose from 5 to 75 mg/mL decreased weekly aggregation rates from ~2% to 0.1%. These results were corroborated by Sane et al., who found that sugars provided the greatest benefit to long-term stability and histidine only provided a minor protective effect during drying [36]. It is worthwhile to note that in both studies, only positively charged amino acid excipients provided stabilizing effects. The authors suggest that the charged guanidinium group of arginine can form strong hydrogen and π-bonds, while its aliphatic chain can favorably interact with hydrophobic surfaces. While histidine can also form strong hydrogen and π-bonds, it is less hydrophobic than the arginine sidechain, which may explain why its stabilization capacity is lower. Interestingly, both arginine and histidine are partially amorphous during drying, which facilitates molecular level interactions between excipient and protein amino acids. This may further explain their ability to enhance stability.

In another study, Faghihi et al. screened several amino acids in spray-dried IgG formulations at 20% and 50% (w/w). The stability, emitted dose (ED), and FPF of the powders were studied. The authors concluded that phenylalanine and cysteine were the best at reducing aggregation during processing and storage. The authors suggest that phenylalaine stabilizes antibodies via hydrophobic and π-stacking interactions with its planar phenyl sidechain. Interestingly, arginine appeared to denature the antibody more than other amino acids based on a decrease in beta-sheet content in FTIR, which seems to contradict other studies and suggests that excipient selection may be dependent upon antibody identity. Concentration did not impact stability, but higher amino acid concentration led to larger mean particle sizes and lower FPF. Of the samples tested, the cysteine formulation had the highest ED and FPF (95 and 70%, respectively) and generated corrugated raisin-like particles. Phenylalanine had slightly lower ED (93%) and FPF (62%) and led to a porous, dimpled surface [47].

3.2 Surface-Active Excipients

Surfactants are another well-studied excipient class for mAb formulation. These are added primarily under the hypothesis that they can minimize physical stress during drying by excluding protein molecules from the air-water interface of droplets. Faghihi et al. reported that cysteine and polysorbate 20 (PS20) synergistically improved the stability of trehalose/IgG formulations compared to either cysteine or PS20 alone [30]. Maa et al. demonstrated that anti-IgE antibody aggregation during spray drying could be ameliorated by the inclusion of PS20 [31]. Additionally, they reported that increasing amounts of PS20 led to smoother antibody/lactose particles, consistent with the idea that PS20 partitions to the surface of the drying particle, excluding other molecules from the interface. Batens et al. reported that inclusion of PS20 in the formulation significantly reduced aggregates during spray drying [48]. However, these excipients can act as destabilizing plasticizers within amorphous solids after spray-drying, so it is likely best practice to minimize the amounts in order to maintain physical stability. Endogenous surfactants formed in the lung are critical to proper lung function [49], and the impact of additional surfactants from DPI formulations on this system is not currently known.

De-aggregation of powder inside a device to primary particles engineered for flight is a critical process for delivery by DPI. Upon inhalation, individual particles must fully aerosolize and disperse, as aggregates will be too large to reach the deep lung. Dispersibility of primary particles is improved by reducing cohesive forces between adjacent particles. Most literature studies aim to achieve this by some combination of increasing surface roughness (thus reducing contact areas) [50] and/or by reducing the surface energy of the particles [51]. Dispersibility-enhancing excipients can improve aerosol performance of powders by facilitating de-aggregation of the powders as they are inhaled into the lung [52].

The role of L-leucine in improving dispersibility and aerosol properties of spray-dried powders for inhalation has been investigated extensively in the literature [53–55]. A recent review by Alhajj et al. discusses this in detail and summarizes regulatory considerations for this excipient [56]. Briefly, nucleation of L-leucine nanocrystals from a nearly-saturated spray solution occurs during the droplet drying process, resulting in a leucine-enriched surface of the particles. This imparts increased surface roughness, as well as localization of the leucine crystals at the particle's interface. These act in combination to improve the dispersibility of the resulting inhalation powders. A related excipient, tri-leucine, is also a powerful surface-active compound that is reported to improve dispersibility [57].

3.3 *Interaction of Spray-Drying Process and Formulation*

The interplay of formulation and spray drying process parameters is critical to development of a pulmonary powder that meets target product profile specifications for deep-lung deposition. In the case of a biotherapeutic API such as a mAb, formulation and process conditions play an additional role in preventing the aggregation and/or degradation of the fragile active. Proteins are sensitive to temperature exposure, and spray drying uses a heated drying gas to produce powder. This apparent conflict is mostly resolved by the impact of evaporative cooling on the temperature of the droplet during spray drying. During much of the drying process, the droplet's temperature is equal to the wet bulb temperature, rather than the boiling point of the solvent [58]. In this way, temperature exposure can be limited, and API degradation can be avoided for many systems. Additional destabilizing forces include dehydration, shear from atomization, and air-liquid interfacial stress [26]. Process conditions and certain excipients can be used to both minimize degradation and maximize yield during spray-drying, setting the particles up for improved stability.

Several seminal papers from Genentech examined the impacts of process parameters and sugar excipients on mAb DPI critical quality attributes [31, 59, 36, 38, 44]. One recent publication aimed at room temperature bulk storage optimized spray drying conditions for three model monoclonal antibodies in parallel [38]. This study compared bench and lab-scale spray dryers and found that larger-scale dryers created powders with lower water content (~4% vs. ~8% for bench scale dryers) and ascribed this to a ~10-fold longer residence time for droplets in the drying chamber.

They also noted that stainless steel is preferable to the glass components found in some benchtop spray dryers, since static buildup on glass can lead to particle accumulation on surfaces. The authors state that particle accumulation on surfaces can also occur via tackiness, which is a complex attribute influenced by glass transition temperature of the particle matrix, formulation, moisture level, and temperature. For instance, they reported that an antibody:trehalose ratio of 1:2 created tacky particles and poor collection yields, but ratios of 2:1 or 1:1 resulted in excellent yields of nontacky powders. Interestingly, they also found that reconstituted spray-dried powders exhibited substantially less turbidity and protein aggregation than corresponding freeze dried formulations, though both had more than the antibody solution prior to drying [38].

Brunaugh et al. employed an anti-streptavidin IgG1 as a model mAb for dry powder inhalation delivery [60]. Spray drying, spray-freeze drying, and lyophilization were compared using a formulation of 58.8% protein, 38.2% sucrose, 1.74% histidine and 0.24% polysorbate 80. A Design of Experiments (DoE) demonstrated optimized atomization conditions that produced inhalation-size particles while limiting aggregation of the mAb. Aerosol performance, secondary structure and aggregation of the dried powders on stability were investigated. A low liquid feed rate, paired with high atomization gas flow rate and low feed concentration, were found to discourage changes in secondary structure and aggregation.

Feasibility of spray drying an IgG1 model antibody for pulmonary delivery was demonstrated by Schüle et al. in 2008 [61]. The formation of aggregates was characterized as a function of spray drying processing conditions and formulation composition. Most formulations were prepared with mannitol as the primary excipient, though inclusion of trehalose was beneficial for protein stabilization, and isoleucine was also beneficial for improving aerodynamic and flow properties. The authors found that reducing the humidity exposure of the formulation during stability was critical to reducing aggregate formation on storage.

3.4 Spray-Drying Scale-Up

A spray-dried inhalation formulation requires special considerations for scale-up from feasibility batches to clinical and commercial scale. These challenges are compounded by the sensitivity of a biotherapeutic API such as a monoclonal antibody.

This section discusses a few of these scale-up challenges and approaches to address them.

Many of the scale-up challenges that apply to pharmaceutical spray drying in general also apply for inhalation formulations of mAbs and other biotherapeutics. Atomization of the feed solution into sufficiently small droplets for inhalation delivery becomes challenging at clinical and commercial scale. A two-fluid atomizer, in which an atomizing gas is used to break up the liquid stream into fine droplets, is commonly used to make particles targeted for pulmonary delivery. A number-up approach, in which the liquid feed is divided between multiple nozzles spraying simultaneously into the dryer, has been used to circumvent these issues [62]. Ultrasonic spray atomizers, such as those found in the Buchi B-90 dryer [63], can produce nearly monodisperse droplets at an appropriate size for inhalation delivery but are not yet scalable to larger dryers. In future work, innovative atomizer geometries should be considered to improve atomization efficiency while minimizing shear exposure for delicate actives. Collection of spray-dried powders is performed using a cyclone separator at large scale. Mosen et al. raised the issue that some cyclones have a particle collection cutoff of ~2 μm, leading to incomplete capture of inhalation-sized particles [64]. In recent years, high-efficiency cyclone geometries have been developed that improved product recovery for <5-micron powders [65].

For conventional spray drying processes, spray solution preparation and handling of protein APIs differs from small molecule APIs. Many protein solutions are supplied frozen, thus, thawing time and appropriate container geometry must be taken into account for larger batch sizes. mAb drug substance is typically prepared in buffer solutions with or without stabilizing excipients. Some amount of solution manipulation is often needed before spray drying to yield the best powder for a dry powder headed for patient dosing. As an example, Table 3 compares an IV formulation with a spray drying solution for bevacizumab. In particular, buffer strength needs to be reduced before spray drying so that the final composition of the spray-dried powder does not contain too much salt for good aerosol performance and to minimize hygroscopicity, which would be challenging for product storage/stability. At a small scale, buffer exchange and excipient introduction can be handled using a traditional dialysis process. However, at a larger scale, the buffer quantities necessary for good dialysis practice (~100x the spray solution volume) become untenable. Tangential flow filtration can solve many of these scaling issues while introducing options for process automation.

Table 3 Example bevacizumab solution compositions for IV drug product and spray drying liquid feed solution

Formulation	Phosphate buffer	mAb concentration	Trehalose concentration	Surfactant concentration	L-leucine concentration
Intravenous solution	50 mM, pH 6.3	30 mg/mL	60 mg/mL	0.04%wt PS 20	None
Spray drying solution	1 mM, pH 6.3	4 mg/mL	4 mg/mL	None	2 mg/mL

Adapted from Shepard et al. [66]

3.5 Device and Packaging

Some challenges for dry powder inhaled mAb formulations include device and packaging considerations. Many antibody-containing dry powder formulations are sensitive to humidity exposure and must therefore be packaged in blisters under controlled conditions. This can be achieved through secondary packaging of capsules in a blister or through powder-filled directly into a blister.

The device selection will be largely driven by stability considerations and optimizing the patient's treatment experience. In order to improve patient compliance and reduce the number of device actuations, developers of pharmaceutical inhalation products are often driven to higher powder doses and higher active loading within the particle. Improvement in the product's delivery efficiency via device design and particle engineering can also help reduce patient burden. This is especially true in the case of mAbs, where the two main delivery options are dry powder inhaler and nebulization. One driver to switch from nebulization to DPI is to reduce the time of dosing from tens of minutes to only seconds. Additional advantages of DPI over nebulizer include device portability, ease of use, and shelf stability for the therapeutic. In the studies discussed here, a formulation with mAb loading of ~50% by weight is feasible, with a typical maximum loading of 30 mg powder per actuation. This corresponds to an approximate maximum mAb dose of 15 mg per DPI actuation. For compounds that require very high pulmonary doses (e.g., >50 mg), nebulization may still be preferred.

Though there is currently a large selection of DPI devices on the market, many were developed with small molecule APIs in mind, particularly using milled crystalline API or an API/lactose-based carrier formulation. The continued evolution and expansion of inhaled biologics treatments will no doubt lead to a focus on developing these new products. After device actuation, material can remain throughout a multi-use device that is exposed to ambient humidity when the device is not in use, which could be problematic for mAb formulations. This has raised the need for either proper cleaning protocols for multi-use DPI devices or the development of single-use devices. These considerations must be examined further as dry powder inhaled mAb products move to later clinical phases.

4 Preclinical and Clinical Case Studies

Since foundational studies in spray-dried mAbs for inhalation performed by Costantino and co-workers on rhuMAbE25 [67], progress has been made in demonstrating preclinical feasibility, efficacy, and clinical safety of dry powder inhaled mAbs prepared by spray drying. While hundreds of studies on pulmonary delivery of spray-dried proteins and other biologics have been conducted, as summarized in numerous review articles [68–72], the literature is more constrained for monoclonal antibodies, perhaps due to the high cost of making and isolating the antibodies themselves. In this section, key preclinical studies of mAbs (and fragments thereof) are highlighted, as well as a detailed discussion of the two exemplary clinical trials completed to date.

Faghihi and colleagues spray-dried infliximab, an anti-TNFα mAb of interest to treat inflammation, for local delivery to the lung via a dry powder formulation [73]. Ten spray-dried formulations were tested with varying ratios of trehalose-to-antibody and cysteine-to-antibody. A lead formulation (approximately 46% trehalose, 16% cysteine, 38% antibody with 0.05% Tween 20) was selected based on exhibiting good aerodynamic properties, low initial aggregation, and low fragmentation on stability. An in vitro cell-based assay demonstrated biological activity was well-preserved in the spray-dried lead formulation compared with the commercial infliximab IV formulation. The anti-inflammatory effect of spray-dried mAb was demonstrated in an allergen-induced mouse model, showing that lung TNFα secretion was significantly reduced when mice were treated with inhaled spray-dried infliximab formulation.

Shepard et al. demonstrated a spray-dried bevacizumab formulation for delivery to the lung as a potential treatment for lung cancer [74]. Bevacizumab, an anti-VEGF compound that can interrupt the formation of new blood vessels in tumors, is approved for IV administration to lung cancer patients, often in combination with chemotherapy. A bevacizumab formulation with trehalose and L-leucine was shown to be physically stable for 24 months at 25 °C as measured by the maintenance of good aerosol properties and in vitro bioactivity [75]. The spray-dried formulation was administered via inhalation in a nude rat model of non-small-cell lung cancer, where it reduced tumor burden both in combination with chemotherapy (IV cisplatin) and on its own. In a follow-up publication, dry powder bevacizumab formulation was paired with cancer-relevant small molecules to produce a combination DPI powder using a simultaneous spray drying process [76].

The first clinical trial of a dry powder inhaled antibody antigen-binding fragment (Fab) was conducted in 2016 on VR942 (NCT02473939) [77]. VR942, also named abrezekimab, is an anti-IL-13 compound indicated for the treatment of asthma that is uncontrolled by corticosteroid therapy. The spray-dried powder was formulated with trehalose, L-leucine, and phosphate buffer as excipients, then administered using a unit-dose DPI device from blisters, filled with either 0.5 mg active or 5.0 mg active. Daily doses of up to 20 mg for 10 days were generally well tolerated by healthy participants and participants with asthma. The authors stated that the results of the study justified further evaluation of the drug in additional clinical trials, though no further clinical trials have been published at the time of this chapter's writing.

Ecleralimab, previously known as CSJ117, is an anti-TSLP antibody fragment from Novartis indicated for asthma treatment. A spray-dried inhalable formulation was formulated with leucine, trileucine, mannitol, and trehalose as excipients. Though no peer-reviewed results were released for this product, clinical trial results are available on Novartis' website. A phase 1 study was completed in 2019, in which participants received 4 mg daily of CSJ117 dry powder for 12 weeks (NCT03138811). Participants with asthma underwent an allergen inhalation challenge, and forced expiratory volume was quantified afterward. Adequate safety and tolerability were demonstrated, as well as an improved response to the allergen inhalation challenge for those treated with CSJ117. Based on these promising results, a phase 2 study is currently recruiting at the time of this chapter's writing (NCT04410523).

5 Outlook and Conclusions

Interest in dry powder inhaled mAbs has grown stronger in recent years as manufacturing technology has progressed, and the first antibody fragment treatments have reached clinical trials. The pipeline of drugs for respiratory illnesses is developing as well, with an increasing fraction of biotherapeutics in the preclinical stages (currently ~30%). Though monoclonal antibodies are the focus of this chapter, many of the same formulation strategies and manufacturing considerations apply to mAb-adjacent actives such as antibody fragments (Fabs) and nanobodies (VHHs) [78]. Here, we discuss our opinions on the outlook for dry powder inhaled mAbs in the treatment of lung disease and ask some open questions for future research.

In the many studies referenced above, a range of model IgG compounds, new chemical entities (both mAb and Fab), and commercially approved mAbs are used to demonstrate formulation and process approaches to DPI development. The generalizability of formulation and manufacturing strategies between different mAbs remains an open question. Due to the structural similarity between mAb compounds, it is more likely that a generalized formulation platform for mAbs is possible than it is for proteins in general. Early-stage screening approaches to determine the sensitivity of mAbs [79] and enzymes [80] have been proposed to address these questions in a materials-sparing fashion.

In other sections of this book, the therapeutic advantages of inhaled delivery of mAbs are discussed more generally. Many of the advantages of a dry powder inhaled mAb formulation relate to stability and patient experience. Dry powder inhalers are also convenient for patients, as they can deliver up to 30 mg powder per actuation in just seconds, compared with minutes-long nebulizer treatments. DPIs can be administered at home with no special equipment, whether once-daily or multiple times a day. Patient education for DPIs is straightforward compared with nebulizers or even metered-dose inhalers, simplifying compliance.

Finally, dry-powder mAb formulations discussed above have often demonstrated excellent physical stability, even at 25 °C. Avoiding cold-chain storage enables simpler distribution to patients with less chance of accidental spoilage, especially in rural areas or developing countries. This could help expand patient access to efficacious treatments for lung diseases. Continuing investigation of dry powder mAb delivery to the lung is warranted to realize these promising benefits for patients in need.

References

1. de Boer AH, et al. Dry powder inhalation: past, present and future. Expert Opin Drug Deliv. 2017:499–512.
2. Hickey AJ. Controlled delivery of inhaled therapeutic agents. J Control Release. 2014:182–8.
3. Lavorini F, Pistolesi M, Usmani OS. Recent advances in capsule-based dry powder inhaler technology. Multidiscip Respir Med. 2017:1–7.
4. Research, C.f.D.E.a. Metered dose inhaler (MDI) and dry powder inhaler (DPI) drug products—quality considerations. FDA, Editor; 2018.
5. Agency EM. Guideline on the pharmaceutical quality of inhalation and nasal products. C.f.m.p.f.h. use, Editor; 2006.
6. Vehring R. Pharmaceutical particle engineering via spray drying. Pharm Res. 2008;25(5):999–1022.
7. Ibrahim M, et al. Protein aggregates in inhaled biologics: challenges and considerations. J Pharm Sci. 2023;
8. Chaurasiya B, Zhao Y-Y. Dry powder for pulmonary delivery: a comprehensive review. Pharmaceutics. 2021:31.
9. Vehring R. Pharmaceutical particle engineering via spray drying. Pharm Res. 2008:999–1022.
10. Chow AHL, et al. Particle engineering for pulmonary drug delivery. Pharm Res. 2007:411–37.
11. Mohammad MA, et al. Effect of mechanical denaturation on surface free energy of protein powders. Colloids Surf B: Biointerfaces. 2016:700–6.
12. Emami F, Keihan Shokooh M, Mostafavi Yazdi SJ. Recent progress in drying technologies for improving the stability and delivery efficiency of biopharmaceuticals. J Pharm Investig. 2023;53(1):35–57.
13. Masters K. Spray drying handbook. George Godwin Ltd; 1985.
14. Baumann JM, Adam MS, Wood JD. Engineering advances in spray drying for pharmaceuticals. Ann Rev Chem Biomol Eng. 2021;12(1):217–40.
15. Seville PC, Li H-Y, Learoyd TP. Spray-dried powders for pulmonary. Drug Deliv. 2007;24(4):307–60.
16. Chen L, et al. Amorphous powders for inhalation drug delivery. Adv Drug Deliv Rev. 2016;100:102–15.
17. Lechanteur A, Evrard B. Influence of composition and spray-drying process parameters on carrier-free DPI properties and Behaviors in the lung: a review. Pharmaceutics. 2020;12(1)
18. Son Y-J, Miller DP, Weers JG. Optimizing spray-dried porous particles for high dose delivery with a portable dry powder inhaler. Pharmaceutics. 2021;13(9)
19. Vishali DA, et al. Spray freeze drying: emerging applications in drug delivery. J Control Release. 2019;300:93–101.
20. Farinha S, et al. Spray freeze drying of biologics: a review and applications for inhalation delivery. Pharm Res. 2022;
21. Maa Y-F, et al. Protein inhalation powders: spray drying vs spray freeze drying. Pharm Res. 1999;16(2):249–54.
22. Hufnagel S, et al. Dry powders for inhalation containing monoclonal antibodies made by thin-film freeze-drying. Int J Pharm. 2022:121637.
23. Dao HM, et al. Degradation of lactoferrin caused by droplet atomization process via two-fluid nozzle: the detrimental effect of air–water interfaces. bioRxiv. 2021:2021.12.06.471411.
24. Mutukuri TT, et al. Electrostatic spray drying for monoclonal antibody formulation. Int J Pharm. 2021;607:120942.
25. Lam JKW, Pan HW. Where the periphery matters: protective excipients investigated in dry powder formulations of inhaled biomacromolecules. In: Inhalation Magazine. Whipsnade and Loophole; 2022.
26. Emami F, et al. Drying Technologies for the Stability and Bioavailability of biopharmaceuticals. Pharmaceutics. 2018:131.

27. Mensink MA, et al. How sugars protect proteins in the solid state and during drying (review): mechanisms of stabilization in relation to stress conditions. Eur J Pharm Biopharm. 2017;114:288–95.
28. Matthews AA, Ee PLR, Ge R. Developing inhaled protein therapeutics for lung diseases. Mol Biomed. 2020;1(1):11.
29. Rajagopal K, et al. Trehalose limits fragment antibody aggregation and influences charge variant formation in spray-dried formulations at elevated temperatures. Mol Pharm. 2019:349–58.
30. Faghihi H, Najafabadi AR, Vatanara A. Optimization and characterization of spray-dried IgG formulations: a design of experiment approach. DARU J Pharm Sci. 2017:22.
31. Maa Y-F, et al. The effect of operating and formulation variables on the morphology of spray-dried protein particles. Pharm Dev Technol. 1997:213–23.
32. Gikanga B, et al. Manufacturing of high-concentration monoclonal antibody formulations via spray drying—the road to manufacturing scale. PDA J Pharm Sci Technol. 2015:59–73.
33. Maury M, et al. Spray-drying of proteins: effects of sorbitol and trehalose on aggregation and FT-IR amide I spectrum of an immunoglobulin G. Eur J Pharm Biopharm. 2005:251–61.
34. Ógáin ON, et al. Particle engineering of materials for oral inhalation by dry powder inhalers. I—particles of sugar excipients (trehalose and raffinose) for protein delivery. Int J Pharm. 2011:23–35.
35. Ramezani V, et al. Screening and evaluation of variables in the formation of antibody particles by spray drying. Powder Technol. 2013:341–6.
36. Sane SU, Wong R, Hsu CC. Raman spectroscopic characterization of drying-induced structural changes in a therapeutic antibody: correlating structural changes with long-term stability. J Pharm Sci. 2003:1005–18.
37. Lassner P, Adler M, Lee G. Formation of insoluble particulates in a spray-dried F(ab')2 fragment. J Pharm Sci. 2014:1021–31.
38. Bowen M, Turok R, Maa Y-F. Spray drying of monoclonal antibodies: investigating powder-based biologic drug substance bulk storage. Dry Technol Int J. 2013:1441–50.
39. Ógáin ON, et al. Particle engineering of materials for oral inhalation by dry powder inhalers. I—particles of sugar excipients (trehalose and raffinose) for protein delivery. Int J Pharm. 2011;405(1):23–35.
40. Bowen M, Turok R, Maa Y-F. Spray drying of monoclonal antibodies: investigating powder-based biologic drug substance bulk storage. Dry Technol. 2013;31(13–14):1441–50.
41. Yoshii H, et al. Effects of protein on retention of ADH enzyme activity encapsulated in trehalose matrices by spray drying. J Food Eng. 2008;87(1):34–9.
42. Liao Y-H, et al. Effects of sucrose and trehalose on the preservation of the native structure of spray-dried lysozyme. Pharm Res. 2002;19(12):1847–53.
43. Colaco CALS, et al. Chemistry of protein stabilization by Trehalose, in formulation and delivery of proteins and peptides. Am Chem Soc. 1994:222–40.
44. Andya JD, et al. The effect of formulation excipients on protein stability and aerosol performance of spray-dried powders of a recombinant humanized anti-IgE monoclonal antibody; 1999. p. 350–8.
45. Schule S, et al. Conformational analysis of protein secondary structure during spray-drying of antibody/mannitol formulations. Eur J Pharm Biopharm. 2007:1–9.
46. Massant J, et al. Formulating monoclonal antibodies as powders for reconstitution at high concentration using spray-drying: Trehalose/amino acid combinations as reconstitution time reducing and stability improving formulations. Eur J Pharm Biopharm. 2020:131–42.
47. Faghihi H, et al. The use of amino acids to prepare physically and conformationally stable spray-dried IgG with enhanced aerosol performance. Int J Pharm. 2014:163–71.
48. Batens M, et al. Formulating monoclonal antibodies as powders for reconstitution at high concentration using spray drying: models and pitfalls. Eur J Pharm Biopharm. 2018:407–22.
49. Hidalgo A, Cruz A, Pérez-Gil J. Barrier or carrier? Pulmonary surfactant and drug delivery. Eur J Pharm Biopharm. 2015;95:117–27.

50. Baldelli A, Vehring R. Analysis of cohesion forces between monodisperse microparticles with rough surfaces. Colloids Surf A Physicochem Eng Asp. 2016;506:179–89.
51. Lechuga-Ballesteros D, Hoe S, Maynor BW. Particle engineering technology for inhaled therapies. In: Pharmaceutical inhalation aerosol technology. CRC Press; 2019. p. 349–61.
52. Ordoubadi M, et al. On the particle formation of leucine in spray drying of inhalable microparticles. Int J Pharm. 2021;592:120102.
53. Chen Y, et al. Drug–polymer–water interaction and its implication for the dissolution performance of amorphous solid dispersions. Mol Pharm. 2015;12(2):576–89.
54. Boraey MA, et al. Improvement of the dispersibility of spray-dried budesonide powders using leucine in an ethanol–water cosolvent system. Powder Technol. 2013;236:171–8.
55. Feng AL, et al. Mechanistic models facilitate efficient development of leucine containing microparticles for pulmonary drug delivery. Int J Pharm. 2011;409(1):156–63.
56. Alhajj N, O'Reilly NJ, Cathcart H. Leucine as an excipient in spray dried powder for inhalation. Drug Discov Today. 2021;26(10):2384–96.
57. Ordoubadi M, et al. Trileucine as a dispersibility enhancer of spray-dried inhalable microparticles. J Control Release. 2021;336:522–36.
58. Abdullahi H, Burcham CL, Vetter T. A mechanistic model to predict droplet drying history and particle shell formation in multicomponent systems. Chem Eng Sci. 2020;224:115713.
59. Maa Y-F, et al. Effect of spray drying and subsequent processing conditions on residual moisture content and physical/biochemical stability of protein inhalation powders. Pharm Res. 1998:768–75.
60. Brunaugh AD, et al. Identification of stability constraints in the particle engineering of an inhaled monoclonal antibody dried powder. J Pharm Sci. 2021;
61. Schüle S, et al. Stabilization of IgG1 in spray-dried powders for inhalation. Eur J Pharm Biopharm. 2008;69(3):793–807.
62. Santos JL, et al. Multi-nozzle spray dryer, method for scale-up of spray dried inhalation powders, multi-nozzle apparatus and use of multiple nozzles in a spray dryer. Hovione International: US; 2019.
63. Heng D, et al. The nano spray dryer B-90. Expert Opin Drug Deliv. 2011;8(7):965–72.
64. Mosén K, et al. Particle formation and capture during spray drying of inhalable particles. Pharm Dev Technol. 2005;9(4):409–17.
65. Alves A, Paiva J, Salcedo R. Cyclone optimization including particle clustering. Powder Technol. 2015;272:14–22.
66. Shepard KB, et al. Local treatment of non-small cell lung cancer with a spray-dried bevacizumab formulation. AAPS PharmSciTech. 2021;22(7):230.
67. Costantino HR, et al. Effect of mannitol crystallization on the stability and aerosol performance of a spray-dried pharmaceutical protein, recombinant humanized anti-IgE monoclonal antibody. J Pharm Sci. 1998;87(11):1406–11.
68. Liang W, et al. Pulmonary delivery of biological drugs. Pharmaceutics. 2020:1025.
69. Chang RYK, et al. Dry powder pharmaceutical biologics for inhalation therapy. Adv Drug Deliv Rev. 2021;172:64–79.
70. Sharma A, et al. Innovative drying technologies for biopharmaceuticals. Int J Pharm. 2021;609:121115.
71. Eedara BB, et al. Spray-dried inhalable powder formulations of therapeutic proteins and peptides. AAPS PharmSciTech. 2021;22(5):185.
72. Carrigy N, Vehring RJPIAT. Engineering stable spray-dried biologic powder for inhalation; 2019. p. 291–326.
73. Faghihi H, et al. Respiratory administration of infliximab dry Powder for local suppression of inflammation. AAPS PharmSciTech. 2019;20(3):128.
74. Shepard KB, et al. Pulmonary delivery of spray-dried monoclonal antibody formulations. 2021.
75. Shepard KB, Pluntze AM, Vodak DT. Room temperature 24-month stability of spray-dried bevacizumab formulation for dry powder inhaler. In: Respiratory drug delivery 2023. Antibes, France; 2023.

76. Shepard KB, Pluntze AM, Vodak DT. Simultaneous spray drying for combination dry powder inhaler formulations. Pharmaceutics. 2022;14(6)
77. Burgess G, et al. Randomized study of the safety and pharmacodynamics of inhaled interleukin-13 monoclonal antibody fragment VR942. EBioMedicine. 2018;35:67–75.
78. Van Heeke G, et al. Nanobodies® as inhaled biotherapeutics for lung diseases. Pharmacol Ther. 2017;169:47–56.
79. Landry M, Pluntze A, Shepard K. Abstract 1736: local treatment of lung cancer by inhaled monoclonal antibodies: a pre-feasibility screening approach for spray drying. Cancer Res, 2022; 82(12_Supplement):1736–1736.
80. Fernandes DA, et al. Formulation of spray dried enzymes for dry powder inhalers: an integrated methodology. Int J Pharm. 2022;615:121492.

Antimicrobial Peptides and Proteins for Inhalation

Yuncheng Wang, Rachel Y. K. Chang, Warwick J. Britton, and Hak-Kim Chan

Abstract Antimicrobial peptides and proteins (APPs) are gaining attraction as powerful agents against multidrug-resistant bacteria. The inhalation delivery of APPs has emerged as a potent tool for treating respiratory infections, given the ability to achieve high local drug concentrations in the lungs and minimizing systemic exposure. This chapter reviews recent in vitro and in vivo research on inhaled APPs, focusing on APP modification and formulation strategies to develop stable and inhalable forms of APPs. These strategies include using *d*-enantiomers, dendrimers, and prodrugs as APP modification strategies and applying new nebulizers, spray drying, and hydrogel techniques as formulation strategies. Despite the substantial progress in enhancing the biological activity of APPs, the biological stability of these novel molecules when aerosolized is yet to be further explored. Similarly, the formulation and production of inhalable liquid and powder formulations of APPs remain relatively unexplored. These domains are, however, critical for successful clinical translation and hence deserve further research efforts.

Keywords Antimicrobial peptides and proteins (APPs), multidrug-resistant bacteria · Lung infections · Inhalation therapy

Y. Wang · R. Y. K. Chang · H.-K. Chan (✉)
Advanced Drug Delivery Group, School of Pharmacy, Faculty of Medicine and Health, University of Sydney, Sydney, NSW, Australia
e-mail: kim.chan@sydney.edu.au

W. J. Britton
Centenary Institute and Faculty of Medicine and Health, University of Sydney, Sydney, NSW, Australia

Department of Clinical Immunology, Royal Prince Alfred Hospital, Camperdown, NSW, Australia

J. K. W. Lam, P. C. L. Kwok (eds.), *Respiratory Delivery of Biologics, Nucleic Acids, and Vaccines*, AAPS Introductions in the Pharmaceutical Sciences 8,
https://doi.org/10.1007/978-3-031-47567-2_4

1 Introduction to Antimicrobial Peptides and Proteins Inhalation Therapy

Antimicrobial peptides and proteins (APPs) in this chapter refer to all the peptides and proteins possessing direct and/or indirect inhibitory effects against bacteria. Direct-acting APPs kill bacteria directly, without the need for other chemicals or systems. By contrast, indirect-acting APPs (host antimicrobial capacity enhancer APPs), including some subunit vaccines/antibodies, recombinant cytokines/chemokines, trigger or enhance other antimicrobial defense responses to kill bacteria. Some APPs have both direct and indirect antimicrobial effects. For example, host defense peptides (HDP) are natural peptides that combat bacterial infection though their direct bactericidal properties and/or by influencing the host's immune responses. Some innate defense regulator (IDR) peptides, which are synthetic and short peptide analogs of HDPs, can have both direct and indirect antimicrobial activities [1]. APPs are becoming important therapeutic alternatives due to the increased number of infections caused by multidrug-resistant (MDR) bacteria. There are many APPs available or in development pipelines [2], and several molecules have been approved by the United States Food and Drug Administration (FDA): colistin, gramicidin D, daptomycin, vancomycin, oritavancin, dalbavancin, and telavancin [3]. APPs possess many advantages over conventional antibiotics, including but not limited to, their lower patterns of resistance, higher antimicrobial activity, and numerous available options. However, the use of APPs faces several limitations for treating lung infections. First, systemic therapeutic routes to deliver APPs normally result in low concentrations in the lung. Second, systemic immune responses to some APPs may cause toxicity. Third, APPs are generally harder to stabilize compared with small molecules.

Inhalation therapy, which has been applied in several commercial antibiotic products, including aztreonam, tobramycin solution, and tobramycin powder [4], is being considered for the delivery of APPs to overcome their limitations caused by systemic routes of administration. By using nebulizers, metered dose inhalers (MDIs), or dry powder inhalers (DPIs), liquid or dry powder formulations of APPs (natural or modified) can be delivered to the lung directly (see Fig. 1) [5]. To generate inhalable aerosols, the aerodynamic diameter of droplets or particles containing drugs should be below 5 μm [6].

Inhalation solution and DPI are the two main formulations being studied for inhalation delivery of APPs. Nebulization has been the primary method for aerosolization and delivery of APPs into the respiratory tract. Commercial nebulizers, including air-jet, ultrasonic, and vibrating mesh nebulizers, are commonly used to deliver inhalation solutions. Air-jet nebulizers use compressed air to deliver the aerosols [7]. Vibrating mesh nebulizers produce aerosols by squeezing liquid through a perforated plate that vibrates at a high frequency (e.g., 100–300 kHz). By comparison, ultrasonic nebulizers exert ultrasonic waves directly into the liquid to cause a breakup at the liquid surface to produce the aerosols [8]. Regardless of the type of nebulizers, they will generate various degrees of stress on the APPs, which

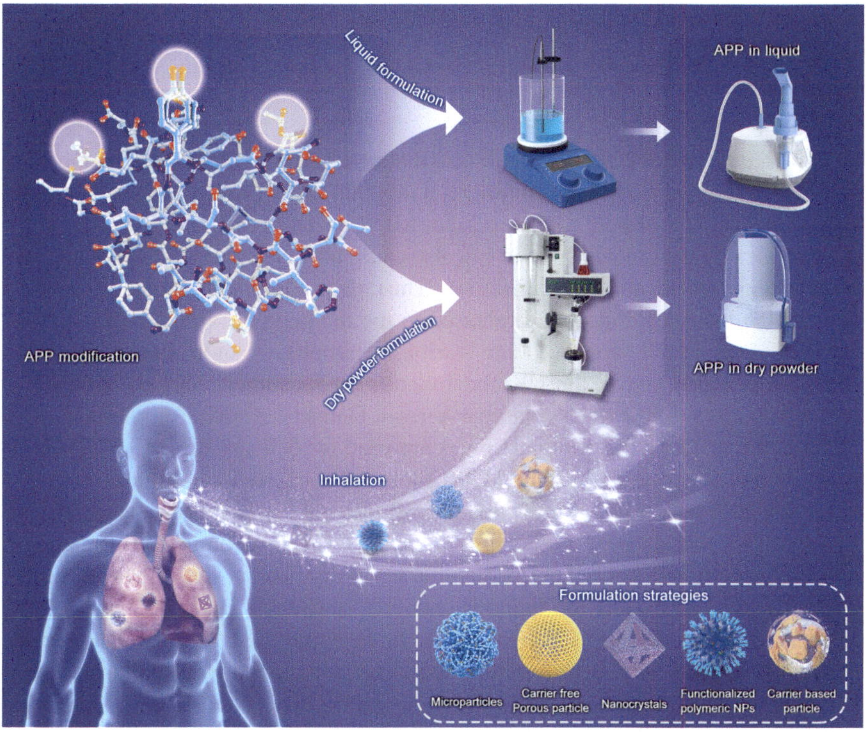

Fig. 1 Production and delivery of APP inhalable formulations (NPs, nanoparticles)

may impact the stability of the biologics. Newer devices, such as the surface acoustic wave atomizer [9], may also be considered for the delivery of APPs.

Multiple studies have been focused on APPs which are stable and inhalable in liquid formulations including peptides (TP359 [10, 11], POL7001 [12, 13], murepavadin [14, 15], LL-37 [16, 17], SPA4 [18, 19] ST3-H2A2 and IL10R1–7 [20, 21]) and proteins (pyocin S2 [22–24], endolysins [25–28], surfactant protein (SP)-D [29–31], flagellin [32] and granulocyte-macrophage colony-stimulating factor [33, 34]). Moreover, different strategies can be used to improve the stability and inhalability of these APP inhalation formulations.

To develop more portable and convenient inhalable products, powder formulation is another option for inhalation drug delivery. Since APPs tend to be more fragile than small molecules, using dry powder formulation eliminates the air-liquid interface and mechanical stresses to the biologics arising from nebulization. However, suitable excipients are necessary to stabilize the APPs in the dry state. In addition, powder formulations involve more complex manufacturing equipment, excipients, and physicochemical characterizations. To date, only a few studies utilized dry powder formulation technologies for APP inhalation delivery. This chapter will focus on strategies applied to the formulation of APPs suitable for inhalation.

2 Commercial and Off-Label Inhaled APPs

2.1 Colistin

Colistin, a cationic polypeptide antibiotic that acts on the bacterial cell membrane, has been developed as the first commercial inhalable APP product. Colistin is commonly used as the last treatment option for MDR bacterial infections because of its serious side effects. Inhaled colistin can also cause adverse effects, including cough, bronchoconstriction, and throat irritation. Thus, its prodrug, colistin methanesulfonate (CMS), was developed, and released to the market as ColiFin® (Approved in Europe), Promixin® (approved in various European Union countries), and Colomycin® (approved in the UK and Ireland) for inhalation. Various nebulizers, such as the Respironics I-neb Adaptive Aerosol Delivery system, Aeroneb Go, Pari eFlow rapid, or Pari LC Sprint nebulizers are used for delivery of CMS [35].

A dry powder formulation of CMS (Colobreathe®, Teva UK Limited) was also developed [36] and approved for inhalational therapy by the European Medicines Agency [37]. A CMS dose of 125 mg in each capsule is administered to patients with the Turbospin delivery device. Another CMS dry powder formulation containing 25 mg of micronized CMS blended with lactose (60–103 μm) was approved for compassionate use with Twincer inhaler in Netherlands [38]. Lactose was used as an excipient as it is FDA-approved and widely used in inhalation formulations [39].

Diverse animal models, such as mouse, rat, piglet, sheep, and monkey, have also been extensively explored to study the pharmacokinetics, safety, and efficacy of colistin [40–47].

2.2 Vancomycin

Vancomycin, a branched tricyclic glycopeptide isolated from *Amycolatopsis orientalis*, has been studied for inhalation delivery in both preclinical and clinical trials [48, 49]. In another clinical study, nebulized vancomycin eradicated methicillin-resistant *Staphylococcus aureus* (MRSA) in 84.3% of the patients [50]. However, the addition of inhaled vancomycin did not affect the rates of MRSA eradication at the end of a 28-day oral antibiotic treatment regimen in a double-blind randomized, placebo-controlled study [51]. Vancomycin dry powder formulation (AeroVanc™) was tested in multiple clinical trials without disclosing its manufacturing technique. AeroVanc was found to be well tolerated, and remained above the usual minimum inhibitory concentration (MIC) values for MRSA for up to 24 h in the sputum [38], and it significantly reduced MRSA sputum density in the subsequent clinical trial [39]. However, the bacterial killing effect was not promising, which was probably why a phase 3 study of AeroVanc in CF (cystic fibrosis) patients did not meet its primary endpoint of improvement in FEV_1 (i.e., lung function) and reduction in pulmonary exacerbations, which made the company halt the project [40]. Although different clinical trials have shown different efficacy results, inhaled vancomycin by nebulization of the intravenous formulation has been applied for off-label use to treat MRSA in CF patients based on its promising safety profile [51].

3 Clinical Trial Stage Products

3.1 Lactoferrin

Lactoferrin is an iron-binding glycoprotein with antimicrobial activity. A phase 1a clinical trial of ALX-009 (a combination of bovine lactoferrin and hypothiocyanite) tested its safety and tolerability. Based on this safety profile, CF and bronchiectasis patients have been recruited for a phase 1b clinical trial to test its efficacy [52]. In a previous study, ALX-009 showed significantly higher antibacterial activity than tobramycin against *Pseudomonas aeruginosa* in CF sputum samples [53]. In a *P. aeruginosa* lung infection model, delivery of aerosolized lactoferrin (200 μg/mouse) reduced both the infiltration of leukocytes and pulmonary bacterial load in CF mice. Furthermore, lactoferrin treatment decreased pulmonary iron levels caused by CF in mice (iron is essential for *P. aeruginosa* to develop antibiotic resistance) [54, 55].

4 Research Stage Studies

Apart from colistin and vancomycin (off-label use), most of the APPs are not available as inhaled therapeutics in the market due to some technical hurdles. Firstly, APP denaturation and degradation in the sputum and lungs, particularly in infected patients, are unpredictable prior to the conduct of in vivo studies. Secondly, APPs are normally more fragile to mechanical stress compared to small molecules, and this increases the uncertainty in both formulation and drug delivery. Thirdly, these biologics may trigger immune responses in humans. Innovative APP modification and formulation strategies are investigated and summarized in this section (see Table 1).

4.1 Strategies in Stabilizing and Enhancing Antimicrobial Peptides and Proteins for Inhalation

Compared with small molecules, APPs require milder conditions to maintain their molecular structures. Owing to the diversity of peptide and protein structures, different strategies were employed for different APPs (Table 1).

Table 1 Summary of research studies on inhaled APPs

Strategy	APP	Formulation	Target bacteria or disease	Key findings	Reference
Enantiomeric optimization	WLBU2	Liquid	*P. aeruginosa*	Enantiomeric optimization of WLBU2 (D8-WLBU2) showed higher stability, higher therapeutic index, and reduced toxicity	[56]
	D8-WLBU2				
Dendrimer synthesis	SET-M33DIM	Liquid	*P. aeruginosa*	Two-branched dimeric form of peptide SET-M33 (SET-M33DIM) was resistant to loss of activity in biological fluids and maintained its antimicrobial activity	[57–59]
Use of prodrug peptides	Prodrug D-Bac8c	Liquid	*P. aeruginosa*	Three prodrugs were cleaved by purified neutrophil elastase to be activated and exert bactericidal activity against *P. aeruginosa* with lower cytotoxicity	[63, 64]
	Prodrug D-HB43				
	Prodrug D-P18				
	Prodrug D-WMR	Liquid	Cystic fibrosis	Prodrug D-WMR inhalation solution showed promising aerosol performance and retained its antimicrobial activity after nebulization	
Peptide sequence optimization	Alpha4-short	Liquid	*P. aeruginosa*	The potency of alpha4-short against *P. aeruginosa* increased with inflammatory cytokines markedly reduced in a murine lung infection study compared to the original alpha4 peptide	[65, 66]
	Pyocin S2, Pyocin SD2	Liquid	*P. aeruginosa*	Intranasal delivery of pyocin SD2 reduced the mortality of *P. aeruginosa*-infected mice while the original pyocin S2 did not decrease the mortality	[22–24]

Use of host antimicrobial capacity enhancer APPs	IDR-HH2 IDR-1018 IDR-1002	Liquid	*Mycobacterium tuberculosis*	Intratracheal instillation of IDR-HH2 and IDR-1018 peptide in solution suppressed the growth of *Mycobacterium tuberculosis* and decreased lung inflammation in a murine tuberculosis infection model, while IDR-1002 treatment did not reduce the bacillary loads during treatment	[69]
	LL-37	Liquid	*P. aeruginosa*	Intranasal delivery of LL-37 solution in a murine lung infection model showed a significant clearance of *P. aeruginosa* from the lungs	[17]
	SPA4	Liquid	*P. aeruginosa*	Intratracheal treatment of SPA4 solution increased phagocytic uptake and entry of *P. aeruginosa* into phagolysosomes, leading to a significant reduction of pulmonary bacterial load	[18, 19]
	ST3-H2A2 IL10R1–7	Liquid	*M. tuberculosis*	Intratracheal delivery of ST3-H2A2 or IL10R1–7 solution in *M. tuberculosis*-infected mice led to dose-dependent increased antimicrobial effect	[20]
	Human surfactant protein D(SP-D) protein	Liquid	*Aspergillus fumigatus*	In a murine aspergillosis lung infection model, intranasal delivery of SP-D solution decreased the mortality rate from 100% to 30% and enhanced protective immunity to *Nippostrongylus. Brasiliensis* in another study	[29, 70]
	PA01 Flagellin	Liquid	*P. aeruginosa*	Intranasal delivery of purified *P. aeruginosa* PA01 flagellin improved lung bacterial clearance with a 90% reduction in mortality in a murine *P. aeruginosa* lung infection model	[32]
	Recombinant murine Granulocyte/ Macrophage Colony Stimulating	Liquid	*S. pneumoniae*	Intratracheal treatment with adenoviral recombinant GM-CSF (Ad-GM-CSF) 3 days prior to *S. pneumoniae* murine lung infection accelerated bacterial clearance by approximately 2 $\log_{10}$	[34]

(continued)

Table 1 (continued)

Strategy	APP	Formulation	Target bacteria or disease	Key findings	Reference
New nebulizers for biologics	RR-11	Liquid	*Mycobacterium smegmatis*	Surface acoustic wave (SAW) nebulizer could generate inhalable aerosols, and maintain the antimicrobial activities of the peptides	[76, 77]
	RY-11				
	LK-8				
	MM-10				
	II-10				
	WW-10				
Spray drying and freeze drying	Lysozyme	Liquid/ powder	*P. aeruginosa*	Different nebulization techniques had different impacts on the stability of lysozyme in solution. Inhalation properties of lysozyme powders were improved by spray drying and freeze-drying	[30, 79–81]
Hydrogel	Lysozyme	Liquid	*P. aeruginosa*	Hydrogel microspheres were produced to maintain bioavailability by protecting lysozyme against uptake by phagocytic cells in the lungs	[82]
Liposomal formulation	Colistin and CMS	Liquid/powder	*P. aeruginosa*	Colistin liposomes (positively charged) remained stable over seven days	[78, 83]
Nanoparticles	SET-M33	Nanoparticles	*P. aeruginosa*	The powder formulations had high dispersibility (fine particle fraction of up to 68%) and disintegrated rapidly into original nanocomplex, with a low moisture content (<5%) and a high Tg	[84, 85]
Mesoporous silica particles	NZX	Mesoporous silica particles	*M. tuberculosis*	NZX MSPs possessed a drug loading rate of 17% wt, and a significantly increased bactericidal activity (>20-fold) than the free NZX	[86]

4.1.1 Enantiomeric Optimization

Enantiomeric optimization could potentially change the properties of the original peptides, which may increase the chance of commercialization of the inhaled drug. This is illustrated in the case of the WLBU2 peptide, which lost its antimicrobial activity after 4 hours of incubation in the presence of epithelial secretions, while its d-enantiomer D8 WLBU-2 remained stable [56]. D8 WLBU2 also showed reduced toxicity and a higher therapeutic index (i.e., the dose of drug that causes therapeutic effects relative to the dose that causes adverse effects) [56].

4.1.2 Dendrimer Synthesis

Dendrimer synthesis is another way of optimizing antimicrobial peptides to be more stable in the presence of proteases [57]. For example, two-branched dimeric form of the cationic antimicrobial peptide SET-M33 (SET-M33DIM) has been produced [58], and it was resistant to a loss of activity in biological fluids and maintained its antimicrobial activity against drug-resistant isolates in vitro and in vivo [57, 59]. According to other studies using various peptides, it has been speculated that the enhanced stability of the peptide might be due to its branched multimeric structure [60] and low amino acid frequency between two branching points of the peptide [61]. Falciani et al. found the occurrence of enhanced stability in enzyme proteolysis with the branched multimeric peptides [60]. Sommer et al. showed that peptides with only one amino acid between two branching points had significantly higher proteolytic stability than those with two or three amino acids [61].

4.1.3 Prodrug Peptides

The use of prodrug peptides could potentially minimize lung protease-induced degradation, enhancing the safety and stability of the antimicrobial peptides.

Forde et al. identified a prodrug of a 13-residue peptide D-WMR that has improved salt resistance with enhanced bactericidal activity in bronchoalveolar lavage fluid (BALF) from CF patients, as well as lower immunogenicity and cytotoxicity [62]. In a subsequent study, pro-WMR, with an aerosol fine particle fraction (i.e., aerosols <5 μm) of 67% after aerosolization using an Aerogen Solo vibrating mesh nebulizer, retained its biological activity. Further, simulated breathing experiments confirmed that a large proportion of the nominal peptide was expected to reach the deep lungs [63]. In another study, three pro-peptides, D-Bac8c$^{2,5\,\mathrm{Leu}}$, D-HB43, and D-P18^{8,} were cleaved by purified neutrophil elastase to be activated and exert bactericidal activity against *P. aeruginosa* with lower cytotoxicity. However, the pro-peptides could only be activated by bronchial alveolar lavage in the presence of sodium chloride [64]. These studies show the value of assessing prodrug peptides for inhalational use to improve their therapeutic properties.

4.1.4 Peptide Sequence Optimization

Sequence modification can affect the function of APPs. For example, alpha4-short is a sequence-optimized (with a reduction in length and increase in the number of positively charged amino acids) peptide from the synthetic short-palate lung and nasal epithelial clone 1 (SPLUNC1)-derived alpha4 peptide [65, 66]. The potency of alpha4-short against *P. aeruginosa* was increased compared to the original alpha4 peptide [66]. Intratracheal delivery of alpha4-short solution significantly decreased bacterial burden in BALF and lung homogenate in a murine *P. aeruginosa* lung infection model. In addition, inflammatory cytokines were markedly reduced compared to mice treated with saline or alpha4, and the host toxicity of alpha4-short, including hemolytic activity and leukopenia, was negligible at concentrations up to 64 μM [66].

Another example is pyocin. Pyocins are produced by *P. aeruginosa* strains, which showed antibacterial activities for intraspecies competition [23]. In one study, intranasal delivery of pyocin S2 solution did not reduce mortality of infected mice 6 h after pulmonary inoculation of *P. aeruginosa*, with approximately 10^5 CFU of *P. aeruginosa* remaining in the lungs. In contrast, intranasal delivery of 75 μg of pyocin SD2 (N-terminal domains homologous with pyocin S2 and C-terminal cytotoxic domain homologous with the tRNase domain of colicin D) in solution resulted in survival of *P. aeruginosa*-infected mice and reduction in bacterial load to only 5 CFU/lung. In addition, the recovered colonies showed no resistance to pyocin SD2 [22, 24]. Sequence optimization could potentially be applied to enhance the antimicrobial activities of APPs.

4.1.5 Use of Host Antimicrobial Capacity Enhancer APPs

Some antimicrobial capacity enhancer APPs target the pathogen indirectly by stimulating the innate host defense system [67]. Most of these APPs are still at the early stage of development as inhaled therapeutics for treating pulmonary infections. Innate defense regulator (IDR) peptides induce chemokine secretion in human peripheral blood mononuclear cells [68]. In one study, intratracheal instillation of IDR-HH2 or IDR-1018 solution suppressed the growth of *Mycobacterium tuberculosis* and decreased lung inflammation in a murine tuberculosis infection model, while IDR-1002 treatment did not reduce the bacillary loads [69]. In another study, LL-37, a human cathelicidin-related 37 amino acid cationic antimicrobial peptide with negligible antimicrobial activity against *P. aeruginosa* in vitro, was administered intranasally in a murine lung infection model. Significant clearance of *P. aeruginosa* in the lungs was observed with enhanced early neutrophil response [16, 17]. Another example is SPA4, which is a surfactant protein A-derived peptide possessing the function of alleviating lung infection and inflammation [18]. In a murine *P. aeruginosa* lung infection model, SPA4 delivered by the intratracheal route promoted phagocytic uptake of *P. aeruginosa* into phagolysosomes, leading to a significant reduction of pulmonary bacterial load [18]. ST3-H2A2 and IL10R1–7 peptides have been designed to inactivate the IL-10-STAT3 pathway as persistent activation of this pathway decreases the antibacterial capacity of the immune cells against *M. tuberculosis*. One study showed that intratracheal delivery of ST3-H2A2 or IL10R1–7 dose-dependently increased the bactericidal effect of the immune cells in *M. tuberculosis*-infected mice [20]. Increased pulmonary activities of nitric oxide synthase, nicotinamide adenine dinucleotide phosphate (NADPH) oxidase, and lysozyme enzymes, as well as lower arginase enzyme activity in the lungs, were also detected.

Aside from the above peptides, some host antimicrobial capacity enhancer proteins have also been developed for inhalation drug studies. Intranasal delivery of Surfactant protein D, an important innate pulmonary immune protein against bacteria, resulted in a decrease of mortality rate from 100% to 30% compared to the control group on day 7 post-infection in a murine aspergillosis lung infection model, and enhanced protective immunity to *Nippostrongylus. Brasiliensis* in another study [29, 70]. Another protein, flagellin, which is a structural protein of bacteria flagella that can activate protective immune responses in lungs, was also studied for treating lung infections [32]. Yu et al. observed strong protective innate responses against *P. aeruginosa* lung infection in mice after intranasal delivery of purified *P. aeruginosa* (strain PA01) flagellin, which consequently improved lung bacterial clearance by >3 $\log_{10}$ by 24 h after infection, with a 90% reduction in mortality 4 days after infection, as compared with untreated mice [32]. Adenoviral recombinant granulocyte-macrophage colony-stimulating factor (Ad-GM-CSF) delivered by intratracheal route also stimulated protective pulmonary immune responses and accelerated bacterial clearance by approximately 2 $\log_{10}$ in a mouse model of *S. pneumoniae* lung infection, reducing the mortality rate from 90% to 10% [34].

4.1.6 Inhalable Combination Formulations

Combination therapy is widely used to develop antimicrobial drugs with potential synergistic effects in killing bacteria.

Lindsay et al. observed dose-dependent synergy of lactoferrin with tobramycin or aztreonam in inhibiting the creation of *P. aeruginosa* (strain PA01) biofilm and in reducing biofilm development on cystic fibrosis airway epithelial cells, while lactoferrin treatment alone presented no killing effect on existing biofilms [71]. Dry powder formulations of lactoferrin and its combination with tobramycin or gentamicin developed by the same group showed promising dispersibility with fine particle fraction ranging from 35% to 42% [72]. Furthermore, the bactericidal effect of the combination formulations of either antibiotic with iron-free lactoferrin was improved with a 5 $\log_{10}$ reduction in CFU compared to a 2 $\log_{10}$ reduction with a single antibiotic formulation [72].

Combination of lysozyme and tobramycin showed significant synergy in vitro against *P. aeruginosa* [73]. In a hamster lung infection model, aerosolized lysozyme and tobramycin resulted in a greater reduction in *P. aeruginosa* bacterial load in the lung and BALF, compared to tobramycin alone. The combination also reduced the inflammatory index compared to tobramycin alone [73]. In another study, a co-spray dried powder containing lysozyme and levofloxacin with median mass aerodynamic diameter below 5 μm showed improved antimicrobial efficacy against *S. aureus* in a rat lung infection model compared with untreated controls, while levofloxacin alone showed no significant killing effect [74].

In a murine model of *S. pneumoniae* lung infection, intranasal delivery of flagellin in combination with oral amoxicillin or intraperitoneal injection of trimethoprim-sulfamethoxazole resulted in higher antimicrobial activity than each single drug. This was attributed to increased neutrophil infiltration in the airways [75]. Furthermore, the combination treatment did not increase inflammation compared to single treatments [75].

4.2 Strategies in Improving Aerosol Properties of APP Inhalation Formulations

Aerosol properties, such as particle (droplet) size, dispersibility, fine particle fraction, and hygroscopicity, are critical for inhalable APP formulations, and these can be optimized to improve the delivery efficiency for APP inhalable formulations.

4.2.1 New Nebulizers for Biologics

A novel surface acoustic wave (SAW) nebulizer which operates at higher frequencies (10–100 MHz vs. 10 kHz–1 MHz) with significantly lower power input than the current commercial ultrasonic nebulizers, was found to reduce the risk of denaturation of biologics by decreasing hydrodynamic shear and cavitation molecular damage [76, 77]. RR-11, RY-11, LK-8, MM-10, II-10, and WW-10 are thymopentin-based cationic peptides that target *Mycobacterium smegmatis*, and the MICs (minimum inhibitory concentrations) of the peptides remained unaltered after SAW nebulization [77]. A Next Generation Cascade Impactor (NGI) test showed that approximately 70% of the aerosols containing WW-10 peptide ranged from 0.98 to 5.39 μm in aerodynamic diameter (i.e., suitable for inhalation) while retaining its antibacterial activity against *M. smegmatis* [77]. This study illustrates the potential of the SAW nebulizer for delivering APPs for the treatment of respiratory diseases.

4.2.2 Spray Drying and Spray Freeze Drying

Spray drying and spray freeze drying are commonly used to produce inhalable powder formulations containing APPs.

In one study, liposomes containing colistin and ciprofloxacin were spray-freeze dried [78]. Excipients, including mannitol, sucrose, and leucine, were used to stabilize colistin and enhance powder dispersibility. The encapsulation efficiencies were > 45%

for both drugs, and the formulation showed higher anti-*Pseudomonal* activity in vitro as compared to each antibiotic alone. Dry powder formulations of lysozyme were also developed to improve the physiochemical stability of the protein. By using d-mannitol and sucrose as excipients, both spray drying (SD) or spray-freeze drying (SFD) methods did not cause lysozyme aggregation [79]. SFD particles possessed larger particle sizes with low density, high rugosity, and high surface area than SD particles. SD and SFD powders showed promising respirable fractions (<5 μm aerodynamic diameter) of 46.7% and 52.1%, respectively [79]. Owing to their large surface area, SFD particles showed greater moisture uptake, and this is deemed less suitable for producing inhalable lysozyme powders because of the higher tendency of powder agglomeration and recrystallization. A co-solvent system was evaluated for spray-dried formulation of lysozyme [80], showing that a higher proportion of ethanol in the feed solution resulted in more wrinkled particles and improved aerosol performance. Furthermore, the presence of trehalose, Tween 20, or phosphate-buffered saline increased lysozyme bioactivity in the spray-dried powders by 5–10% compared with those without these excipients. Also, lower inlet temperature (80 °C compared to 130 °C) and shorter storage time (24 h compared to 72 h) of the feed solution reduced the activity loss of the powders [80], indicating the vulnerability of lysozyme to high temperature and organic solvent. In another study, three factors (density, total solid content, and percentage of sucrose and mannitol) were tested to optimize the spray-freeze dried lysozyme formulations [81]. Higher protein concentration, lower sugar amount, and a lower total solid content in the feed suspension resulted in improved aerosol performance of the powders. The powders remained stable after 6 months of storage at −20 °C [81].

4.2.3 Hydrogel

Hydrogel microspheres were produced to retain the bioavailability by protecting lysozyme against uptake by phagocytic cells in the lung [82]. Polyethylene glycol (PEG)-based microgels loaded with lysozyme showed an encapsulation efficiency of 51.5%. In addition, particle size could be controlled by optimizing the ratio of PEG: dextran in the aqueous two-phase system. Microgels released the loaded protein over a 6–14 days period, depending on the macromonomer type. Further MTT assay revealed that the particles did not show significant cytotoxicity, and their recognition by alveolar macrophages was significantly lower than that for polystyrene control particles [82].

4.2.4 Liposomal Formulation

Wallace et al. explored liposomal formulations for pulmonary delivery of colistin and CMS to achieve constant release [83]. Liposomes with a size of 160–190 nm and encapsulation efficiency of 50% for colistin were produced. CMS liposomes were stable for 48 h, with an increase in particle size and a reversal in zeta potential from negative to positive, while colistin liposomes (positively charged) remained stable over seven days [83].

4.2.5 Nanoparticles

SET-M33 nanoparticles (M33-NS) were produced by incorporating SET-M33 on single-chain dextran, resulting in sustained antimicrobial activity against *P. aeruginosa* in vitro. Cytotoxicity of the formulation was relatively low toward different human and murine cell lines [84]. Biodistribution and pharmacokinetics of labeled SET-M33 and M33-NS were assessed in healthy rats after intratracheal treatment. PET-CT image showed localized delivery of the two compounds in the lungs, with M33-NS having a much longer pulmonary residence time of 12.6 h. Both compounds had similar antimicrobial activities in a *P. aeruginosa* lung infection mouse model, exhibiting a complete eradication at a dose of 5 mg/kg [84].

Polymeric nanoparticles (PEG and miktoarm polymers) were also explored for lysozyme, and co-spray dried with excipients (mannitol, trehalose, and leucine) after encapsulation [85]. The powder formulations had high dispersibility (fine particle fraction of up to 68%) and disintegrated rapidly into the original nanocomplex, with a low moisture content (<5%) and a high Tg (>125 °C) [85].

4.2.6 Mesoporous Silica Particles

As *M. tuberculosis* primarily resides inside alveolar macrophages, mesoporous silica particles (MSPs) with strong negative surface charge were used to enable intracellular delivery of cationic anti-mycobacterial peptide NZX. In an intracellular infection model, the developed NZX MSPs possessed a drug loading rate of 17% wt and a significantly increased bactericidal activity (>20-fold) than the free NZX. Moreover, intratracheal delivery of NZX MSPs (once daily for 5 days) reduced *M. tuberculosis* bacterial load by 88% (0.9 $\log_{10}$) in a murine lung infection model [86]. The MSP formulation could potentially be used to target other intracellular pathogens. In one study, intravenous injections of MSPs in female and male mice were found to be well tolerated by the animals, and the degradation and clearance of the MSPs were detected with no significant lesion observed [87]. However, the use of silica nanomaterials as carriers for drug delivery in the lungs is still in its early stages. Its safety and tolerability should be investigated in the future.

5 Future Direction and Perspective

Inhaled APP therapy is showing considerable clinical promise for the treatment of MDR lung infections. However, most of the APPs are still in the early stage of development due to various technical hurdles [69].

Modification of APPs, including the use of enantiomeric optimization, dendritic form, or prodrugs, are potential avenues for investigation to formulate stable APPs for inhalation. In addition, formulation strategies for inhalation drugs, such as nanoparticle technology, can be applied to enhance the stability of APPs. Combination strategies, such as the combination of innate host defense peptides/proteins and antimicrobial biologics, are also worth consideration to enhance the efficacy of the APPs.

According to the current examples, similar strategies have been applied to formulate and deliver peptides and proteins. Because the development of inhaled APPs is still at its early stage, more studies in the future are required to confirm any specific differences between inhaled peptides and inhaled proteins in various aspects such as formulation design, drug delivery, or pharmacokinetics.

To overcome the technical challenges in this area, more investigations are required in the future.

Aside from these technical issues, inhaled APP product development is also a business decision that requires investments and collaborations in addition to resolving the technological barriers. In the APP manufacturing process, the potential high cost of APP production is a consideration. In the formulation process, ensuring aerosol performance while retaining the biological and physicochemical stability of the APPs requires repeated assessments. In the clinical trial stage, these APPs may stimulate immune responses and potential toxicity in humans, which would require extra safety assessments. The above issues will potentially translate to a higher cost for developing inhaled APP products. To overcome the economic hurdle above, close collaboration among different participants is needed to ensure the continued discovery and development of APP inhalable formulations, leading to large-scale production. Nevertheless, despite these challenges, inhaled APP therapy still holds tremendous promise for the therapy of respiratory infections from MDR pathogens.

Acknowledgments We acknowledge support from the Australian National Health and Medical Research Council to H.-K. C. (APP1140617) and W. J. B. (APP1153493) and from the National Institutes of Health to H.-K. C.

References

1. Brunet J, Eichner C, Male R. The FTZ-F1 gene encodes two functionally distinct nuclear receptor isoforms in the ectoparasitic copepod salmon louse (Lepeophtheirus salmonis). PLoS One. 2021;16:e0251575. https://doi.org/10.1371/journal.pone.0251575.
2. Wang G, Li X, Wang Z. APD3: the antimicrobial peptide database as a tool for research and education. Nucleic Acids Res. 2016a;44:D1087–93. https://doi.org/10.1093/nar/gkv1278.
3. Patrulea V, Borchard G, Jordan O. An update on antimicrobial peptides (AMPs) and their delivery strategies for wound infections. Pharmaceutics. 2020;12:840. https://doi.org/10.3390/pharmaceutics12090840.
4. Daniels LM, Juliano J, Marx A, Weber DJ. Inhaled antibiotics for hospital-acquired and ventilator-associated pneumonia. Clin Infect Dis. 2017;64:386–7. https://doi.org/10.1093/cid/ciw726.
5. Zhou QT, Leung SS, Tang P, Parumasivam T, Loh ZH, Chan HK. Inhaled formulations and pulmonary drug delivery systems for respiratory infections. Adv Drug Deliv Rev. 2015;85:83–99. https://doi.org/10.1016/j.addr.2014.10.022.
6. Mortensen NP, Hickey AJ. Targeting inhaled therapy beyond the lungs. Respiration. 2014;88:353–4. https://doi.org/10.1159/000367852.

7. Hertel SP, Winter G, Friess W. Protein stability in pulmonary drug delivery via nebulization. Adv Drug Deliv Rev. 2015;93:79–94. https://doi.org/10.1016/j.addr.2014.10.003.
8. Cipolla D, Clark A, Chan H-K, Gonda I, Shire S. Assessment of aerosol delivery systems for recombinant human deoxyribonuclease. STP Pharm Sci. 1994;4:50–62.
9. Cortez-Jugo C, Masoumi S, Chan PPY, Friend J, Yeo L. Nebulization of siRNA for inhalation therapy based on a microfluidic surface acoustic wave platform. Ultrason Sonochem. 2022;88:106088. https://doi.org/10.1016/j.ultsonch.2022.106088.
10. Dosunmu EF, Chaudhari AA, Bawage S, Bakeer MK, Owen DR, Singh SR, et al. Novel cationic peptide TP359 down-regulates the expression of outer membrane biogenesis genes in Pseudomonas aeruginosa: a potential TP359 anti-microbial mechanism. BMC Microbiol. 2016;16:192. https://doi.org/10.1186/s12866-016-0808-2.
11. Dosunmu EF, Emeh RO, Dixit S, Bakeer MK, Coats MT, Owen DR, et al. The anti-microbial peptide TP359 attenuates inflammation in human lung cells infected with Pseudomonas aeruginosa via TLR5 and MAPK pathways. PLoS One. 2017;12:e0176640. https://doi.org/10.1371/journal.pone.0176640.
12. Srinivas N, Jetter P, Ueberbacher BJ, Werneburg M, Zerbe K, Steinmann J, et al. Peptidomimetic antibiotics target outer-membrane biogenesis in Pseudomonas aeruginosa. Science. 2010;327:1010–3. https://doi.org/10.1126/science.1182749.
13. Cigana C, Bernardini F, Facchini M, Alcala-Franco B, Riva C, De Fino I, et al. Efficacy of the novel antibiotic POL7001 in preclinical models of Pseudomonas aeruginosa pneumonia. Antimicrob Agents Chemother. 2016;60:4991–5000. https://doi.org/10.1128/AAC.00390-16.
14. Martin-Loeches I, Dale GE, Torres A. Murepavadin: a new antibiotic class in the pipeline. Expert Rev Anti-Infect Ther. 2018;16:259–68. https://doi.org/10.1080/14787210.2018.1441024.
15. Bernardini F, Dale G, Wach A, Obrecht D. WS01-4 pharmacokinetics and pharmacodynamics of murepavadin (POL7080) in neutropenic lung infection models when evaluated by aerosol administration. J Cyst Fibros. 2019;18:S2.
16. Pompilio A, Scocchi M, Pomponio S, Guida F, Di Primio A, Fiscarelli E, et al. Antibacterial and anti-biofilm effects of cathelicidin peptides against pathogens isolated from cystic fibrosis patients. Peptides. 2011;32:1807–14. https://doi.org/10.1016/j.peptides.2011.08.002.
17. Beaumont PE, Mchugh B, Gwyer Findlay E, Mackellar A, Mackenzie KJ, Gallo RL, et al. Cathelicidin host defence peptide augments clearance of pulmonary Pseudomonas aeruginosa infection by its influence on neutrophil function in vivo. PLoS One. 2014;9:e99029. https://doi.org/10.1371/journal.pone.0099029.
18. Awasthi S, Rahman N, Rui B, Kumar G, Awasthi V, Breshears M, et al. Lung and general health effects of toll-like receptor-4 (TLR4)-interacting SPA4 peptide. BMC Pulm Med. 2020;20:179. https://doi.org/10.1186/s12890-020-01187-7.
19. Awasthi S, Singh B, Ramani V, Xie J, Kosanke S. TLR4-interacting SPA4 peptide improves host defense and alleviates tissue injury in a mouse model of Pseudomonas aeruginosa lung infection. PLoS One. 2019;14:e0210979. https://doi.org/10.1371/journal.pone.0210979.
20. Upadhyay R, Sanchez-Hidalgo A, Wilusz CJ, Lenaerts AJ, Arab J, Yeh J, et al. Host directed therapy for chronic tuberculosis via intrapulmonary delivery of aerosolized peptide inhibitors targeting the IL-10-STAT3 pathway. Sci Rep. 2018;8:16610. https://doi.org/10.1038/s41598-018-35023-0.
21. Redford PS, Murray PJ, O'garra A. The role of IL-10 in immune regulation during M. Tuberculosis infection. Mucosal Immunol. 2011;4:261–70. https://doi.org/10.1038/mi.2011.7.
22. Mccaughey LC, Josts I, Grinter R, White P, Byron O, Tucker NP, et al. Discovery, characterization and in vivo activity of pyocin SD2, a protein antibiotic from Pseudomonas aeruginosa. Biochem J. 2016;473:2345–58. https://doi.org/10.1042/BCJ20160470.
23. Michel-Briand Y, Baysse C. The pyocins of Pseudomonas aeruginosa. Biochimie. 2002;84:499–510. https://doi.org/10.1016/s0300-9084(02)01422-0.
24. Ghequire MG, De Mot R. Ribosomally encoded antibacterial proteins and peptides from pseudomonas. FEMS Microbiol Rev. 2014;38:523–68. https://doi.org/10.1111/1574-6976.12079.

25. Doehn JM, Fischer K, Reppe K, Gutbier B, Tschernig T, Hocke AC, et al. Delivery of the endolysin Cpl-1 by inhalation rescues mice with fatal pneumococcal pneumonia. J Antimicrob Chemother. 2013;68:2111–7. https://doi.org/10.1093/jac/dkt131.
26. Raz A, Serrano A, Hernandez A, Euler CW, Fischetti VA. Isolation of phage Lysins that effectively kill Pseudomonas aeruginosa in mouse models of lung and skin infection. Antimicrob Agents Chemother. 2019;63:e00024–19. https://doi.org/10.1128/AAC.00024-19.
27. Bae JY, Jun KI, Kang CK, Song KH, Choe PG, Bang JH, et al. Efficacy of intranasal administration of the recombinant Endolysin SAL200 in a lethal murine Staphylococcus aureus pneumonia model. Antimicrob Agents Chemother. 2019;63:e02009–18. https://doi.org/10.1128/AAC.02009-18.
28. Wang Y, Khanal D, Chang RYK, Shang X, Yang H, Britton WJ, et al. Can bacteriophage endolysins be nebulised for inhalation delivery against Streptococcus pneumoniae? Int J Pharm. 2020;591:119982. https://doi.org/10.1016/j.ijpharm.2020.119982.
29. Thawer S, Auret J, Schnoeller C, Chetty A, Smith K, Darby M, et al. Surfactant protein-D is essential for immunity to helminth infection. PLoS Pathog. 2016;12:e1005461. https://doi.org/10.1371/journal.ppat.1005461.
30. Singh N, Yadav P, Gaur P, Gaur M, Yadav AB. Protein stability and functional activity during nebulization: a comparative study of three nebulizer! bioRxiv; 2020. https://doi.org/10.1101/2020.05.09.085720.
31. Allen MJ, Harbeck R, Smith B, Voelker DR, Mason RJ. Binding of rat and human surfactant proteins a and D to aspergillus fumigatus conidia. Infect Immun. 1999;67:4563–9. https://doi.org/10.1128/IAI.67.9.4563-4569.1999.
32. Yu FS, Cornicelli MD, Kovach MA, Newstead MW, Zeng X, Kumar A, et al. Flagellin stimulates protective lung mucosal immunity: role of cathelicidin-related antimicrobial peptide. J Immunol. 2010;185:1142–9. https://doi.org/10.4049/jimmunol.1000509.
33. Trapnell BC, Whitsett JA. Gm-CSF regulates pulmonary surfactant homeostasis and alveolar macrophage-mediated innate host defense. Annu Rev Physiol. 2002;64:775–802. https://doi.org/10.1146/annurev.physiol.64.090601.113847.
34. Steinwede K, Tempelhof O, Bolte K, Maus R, Bohling J, Ueberberg B, et al. Local delivery of GM-CSF protects mice from lethal pneumococcal pneumonia. J Immunol. 2011;187:5346–56. https://doi.org/10.4049/jimmunol.1101413.
35. Promixin 1 million International Units (IU) Powder for Nebuliser Solution. https://www.medicines.org.uk/emc/product/4/smpc (2020). Accessed 28 Feb 2023.
36. Quon BS, Goss CH, Ramsey BW. Inhaled antibiotics for lower airway infections. Ann Am Thorac Soc. 2014;11:425–34. https://doi.org/10.1513/AnnalsATS.201311-395FR.
37. Kaplan S, Patino O, Rainville C, Madison T. Assessment of colistimethate sodium (COLOBREATHE) risk minimization measures implemented in the European Union: a cross-sectional study. Pharmacoepidemiol Drug Saf. 2020;29:219–23. https://doi.org/10.1002/pds.4940.
38. Westerman EM, De Boer AH, Le Brun PP, Touw DJ, Roldaan AC, Frijlink HW, et al. Dry powder inhalation of colistin in cystic fibrosis patients: a single dose pilot study. J Cyst Fibros. 2007;6:284–92. https://doi.org/10.1016/j.jcf.2006.10.010.
39. Baldrick P, Bamford DG. A toxicological review of lactose to support clinical administration by inhalation. Food Chem Toxicol. 1997;35:719–33. https://doi.org/10.1016/s0278-6915(97)00041-0.
40. Yapa SWS, Li J, Porter CJ, Nation RL, Patel K, Mcintosh MP. Population pharmacokinetics of colistin methanesulfonate in rats: achieving sustained lung concentrations of colistin for targeting respiratory infections. Antimicrob Agents Chemother. 2013;57:5087–95. https://doi.org/10.1128/AAC.01127-13.
41. Marchand S, Bouchene S, De Monte M, Guilleminault L, Montharu J, Cabrera M, et al. Pharmacokinetics of Colistin Methansulphonate (CMS) and Colistin after CMS Nebulisation in Baboon Monkeys. Pharm Res. 2015;32:3403–14. https://doi.org/10.1007/s11095-015-1716-0.

42. Landersdorfer CB, Nguyen TH, Lieu LT, Nguyen G, Bischof RJ, Meeusen EN, et al. Substantial targeting advantage achieved by pulmonary administration of colistin methanesulfonate in a large-animal model. Antimicrob Agents Chemother. 2017;61:e01934–16. https://doi.org/10.1128/AAC.01934-16.
43. Lu Q, Girardi C, Zhang M, Bouhemad B, Louchahi K, Petitjean O, et al. Nebulized and intravenous colistin in experimental pneumonia caused by Pseudomonas aeruginosa. Intensive Care Med. 2010;36:1147–55. https://doi.org/10.1007/s00134-010-1879-4.
44. Marchand S, Gobin P, Brillault J, Baptista S, Adier C, Olivier JC, et al. Aerosol therapy with colistin methanesulfonate: a biopharmaceutical issue illustrated in rats. Antimicrob Agents Chemother. 2010;54:3702–7. https://doi.org/10.1128/AAC.00411-10.
45. Lin YW, Zhou QT, Cheah SE, Zhao J, Chen K, Wang J, et al. Pharmacokinetics/pharmacodynamics of pulmonary delivery of Colistin against Pseudomonas aeruginosa in a mouse lung infection model. Antimicrob Agents Chemother. 2017a;61:e02025–16. https://doi.org/10.1128/AAC.02025-16.
46. Tewes F, Brillault J, Gregoire N, Olivier JC, Lamarche I, Adier C, et al. Comparison between colistin sulfate dry powder and solution for pulmonary delivery. Pharmaceutics. 2020;12:557. https://doi.org/10.3390/pharmaceutics12060557.
47. Lin YW, Zhou QT, Hu Y, Onufrak NJ, Sun S, Wang J, et al. Pulmonary pharmacokinetics of Colistin following Administration of dry Powder Aerosols in rats. Antimicrob Agents Chemother. 2017b;61:e00973–17. https://doi.org/10.1128/AAC.00973-17.
48. Park CW, Li X, Vogt FG, Hayes D Jr, Zwischenberger JB, Park ES, et al. Advanced spray-dried design, physicochemical characterization, and aerosol dispersion performance of vancomycin and clarithromycin multifunctional controlled release particles for targeted respiratory delivery as dry powder inhalation aerosols. Int J Pharm. 2013;455:374–92. https://doi.org/10.1016/j.ijpharm.2013.06.047.
49. Waterer G, Lord J, Hofmann T, Jouhikainen T. Phase I, dose-escalating study of the safety and pharmacokinetics of inhaled dry-powder vancomycin (AeroVanc) in volunteers and patients with cystic fibrosis: a new approach to therapy for methicillin-resistant Staphylococcus aureus. Antimicrob Agents Chemother. 2020;64:e01776–19. https://doi.org/10.1128/AAC.01776-19.
50. Hayes D Jr, Murphy BS, Mullett TW, Feola DJ. Aerosolized vancomycin for the treatment of MRSA after lung transplantation. Respirology. 2010;15:184–6. https://doi.org/10.1111/j.1440-1843.2009.01647.x.
51. Dezube R, Jennings MT, Rykiel M, Diener-West M, Boyle MP, Chmiel JF, et al. Eradication of persistent methicillin-resistant Staphylococcus aureus infection in cystic fibrosis. J Cyst Fibros. 2019;18:357–63. https://doi.org/10.1016/j.jcf.2018.07.005.
52. Dose Escalation Study of ALX-009 in Healthy Men and Cystic Fibrosis (CF) and Non-CF Bronchiectasis Patients. https://ClinicalTrials.gov/show/NCT02598999 (2022). Accessed 28 Feb 2023.
53. Tunney MM, Payne JE, Mcgrath SJ, Einarsson GG, Ingram RJ, Gilpin DF, et al. Activity of hypothiocyanite and lactoferrin (ALX-009) against respiratory cystic fibrosis pathogens in sputum. J Antimicrob Chemother. 2018;73:3391–7. https://doi.org/10.1093/jac/dky357.
54. Cutone A, Lepanto MS, Rosa L, Scotti MJ, Rossi A, Ranucci S, et al. Aerosolized bovine Lactoferrin counteracts infection, inflammation and iron dysbalance in a cystic fibrosis mouse model of pseudomonas aeruginosa chronic lung infection. Int J Mol Sci. 2019;20:2128. https://doi.org/10.3390/ijms20092128.
55. Hazlett HF, Hampton TH, Aridgides DS, Armstrong DA, Dessaint JA, Mellinger DL, et al. Altered iron metabolism in cystic fibrosis macrophages: the impact of CFTR modulators and implications for Pseudomonas aeruginosa survival. Sci Rep. 2020;10:10935. https://doi.org/10.1038/s41598-020-67729-5.
56. Di YP, Lin Q, Chen C, Montelaro RC, Doi Y, Deslouches B. Enhanced therapeutic index of an antimicrobial peptide in mice by increasing safety and activity against multidrug-resistant bacteria. Sci Adv. 2020;6:eaay6817. https://doi.org/10.1126/sciadv.aay6817.

57. Bracci L, Falciani C, Lelli B, Lozzi L, Runci Y, Pini A, et al. Synthetic peptides in the form of dendrimers become resistant to protease activity. J Biol Chem. 2003;278:46590–5. https://doi.org/10.1074/jbc.M308615200.
58. Quercini L, Brunetti J, Riolo G, Bindi S, Scali S, Lampronti I, et al. An antimicrobial molecule mitigates signs of sepsis in vivo and eradicates infections from lung tissue. FASEB J. 2020;34:192–207. https://doi.org/10.1096/fj.201901896RR.
59. Pini A, Falciani C, Mantengoli E, Bindi S, Brunetti J, Iozzi S, et al. A novel tetrabranched antimicrobial peptide that neutralizes bacterial lipopolysaccharide and prevents septic shock in vivo. FASEB J. 2010;24:1015–22. https://doi.org/10.1096/fj.09-145474.
60. Falciani C, Lozzi L, Pini A, Corti F, Fabbrini M, Bernini A, et al. Molecular basis of branched peptides resistance to enzyme proteolysis. Chem Biol Drug Des. 2007;69:216–21. https://doi.org/10.1111/j.1747-0285.2007.00487.x.
61. Sommer P, Fluxa VS, Darbre T, Reymond JL. Proteolysis of peptide dendrimers. Chembiochem. 2009;10:1527–36. https://doi.org/10.1002/cbic.200900060.
62. Forde E, Schutte A, Reeves E, Greene C, Humphreys H, Mall M, et al. Differential in vitro and in vivo toxicities of antimicrobial peptide prodrugs for potential use in cystic fibrosis. Antimicrob Agents Chemother. 2016;60:2813–21. https://doi.org/10.1128/AAC.00157-16.
63. Forde E, Kelly G, Sweeney L, Fitzgerald-Hughes D, Macloughlin R, Devocelle M. Vibrating mesh nebulisation of pro-antimicrobial peptides for use in cystic fibrosis. Pharmaceutics. 2019;11:239. https://doi.org/10.3390/pharmaceutics11050239.
64. Forde E, Humphreys H, Greene CM, Fitzgerald-Hughes D, Devocelle M. Potential of host defense peptide prodrugs as neutrophil elastase-dependent anti-infective agents for cystic fibrosis. Antimicrob Agents Chemother. 2014;58:978–85. https://doi.org/10.1128/AAC.01167-13.
65. Walton WG, Ahmad S, Little MS, Kim CS, Tyrrell J, Lin Q, et al. Structural features essential to the antimicrobial functions of human SPLUNC1. Biochemistry. 2016;55:2979–91. https://doi.org/10.1021/acs.biochem.6b00271.
66. Jiang S, Deslouches B, Chen C, Di ME, Di YP. Antibacterial properties and efficacy of a novel SPLUNC1-derived antimicrobial peptide, alpha4-short, in a murine model of respiratory infection. MBio. 2019;10:e00226–19. https://doi.org/10.1128/mBio.00226-19.
67. Hartl D, Tirouvanziam R, Laval J, Greene CM, Habiel D, Sharma L, et al. Innate immunity of the lung: from basic mechanisms to translational medicine. J Innate Immun. 2018;10:487–501. https://doi.org/10.1159/000487057.
68. Nijnik A, Madera L, Ma S, Waldbrook M, Elliott MR, Easton DM, et al. Synthetic cationic peptide IDR-1002 provides protection against bacterial infections through chemokine induction and enhanced leukocyte recruitment. J Immunol. 2010;184:2539–50. https://doi.org/10.4049/jimmunol.0901813.
69. Rivas-Santiago B, Castaneda-Delgado JE, Rivas Santiago CE, Waldbrook M, Gonzalez-Curiel I, Leon-Contreras JC, et al. Ability of innate defence regulator peptides IDR-1002, IDR-HH2 and IDR-1018 to protect against Mycobacterium tuberculosis infections in animal models. PLoS One. 2013;8:e59119. https://doi.org/10.1371/journal.pone.0059119.
70. Singh M, Madan T, Waters P, Sonar S, Singh SK, Kamran MF, et al. Therapeutic effects of recombinant forms of full-length and truncated human surfactant protein D in a murine model of invasive pulmonary aspergillosis. Mol Immunol. 2009;46:2363–9. https://doi.org/10.1016/j.molimm.2009.03.019.
71. Moreau-Marquis S, Coutermarsh B, Stanton BA. Combination of hypothiocyanite and lactoferrin (ALX-109) enhances the ability of tobramycin and aztreonam to eliminate Pseudomonas aeruginosa biofilms growing on cystic fibrosis airway epithelial cells. J Antimicrob Chemother. 2015;70:160–6. https://doi.org/10.1093/jac/dku357.
72. Marshall LJ, Oguejiofor W, Price R, Shur J. Investigation of the enhanced antimicrobial activity of combination dry powder inhaler formulations of lactoferrin. Int J Pharm. 2016;514:399–406. https://doi.org/10.1016/j.ijpharm.2016.09.034.
73. Bhavsar T, Liu M, Liu X, Cantor J. Aerosolized recombinant human lysozyme enhances the bactericidal effect of tobramycin in a hamster model of Pseudomonas aeruginosa-induced pneumonia. Exp Lung Res. 2011;37:536–41. https://doi.org/10.3109/01902148.2011.609578.

74. Gupta PV, Nirwane AM, Nagarsenker MS. Inhalable levofloxacin liposomes complemented with lysozyme for treatment of pulmonary infection in rats: effective antimicrobial and Antibiofilm strategy. AAPS PharmSciTech. 2018;19:1454–67. https://doi.org/10.1208/s12249-017-0945-4.
75. Matarazzo L, Casilag F, Porte R, Wallet F, Cayet D, Faveeuw C, et al. Therapeutic synergy between antibiotics and pulmonary toll-like receptor 5 stimulation in antibiotic-sensitive or -resistant pneumonia. Front Immunol. 2019;10:723. https://doi.org/10.3389/fimmu.2019.00723.
76. Qi A, Friend JR, Yeo LY, Morton DA, Mcintosh MP, Spiccia L. Miniature inhalation therapy platform using surface acoustic wave microfluidic atomization. Lab Chip. 2009;9:2184–93. https://doi.org/10.1039/b903575c.
77. Wang Y, Rezk AR, Khara JS, Yeo LY, Ee PL. Stability and efficacy of synthetic cationic antimicrobial peptides nebulized using high frequency acoustic waves. Biomicrofluidics. 2016b;10:034115. https://doi.org/10.1063/1.4953548.
78. Yu S, Wang S, Zou P, Chai G, Lin YW, Velkov T, et al. Inhalable liposomal powder formulations for co-delivery of synergistic ciprofloxacin and colistin against multi-drug resistant gram-negative lung infections. Int J Pharm. 2020;575:118915. https://doi.org/10.1016/j.ijpharm.2019.118915.
79. Brunaugh AD, Wu T, Kanapuram SR, Smyth HDC. Effect of particle formation process on characteristics and aerosol performance of Respirable protein powders. Mol Pharm. 2019;16:4165–80. https://doi.org/10.1021/acs.molpharmaceut.9b00496.
80. Ji S, Thulstrup PW, Mu H, Hansen SH, Van De Weert M, Rantanen J, et al. Investigation of factors affecting the stability of lysozyme spray dried from ethanol-water solutions. Int J Pharm. 2017;534:263–71. https://doi.org/10.1016/j.ijpharm.2017.10.021.
81. Ferrati S, Wu T, Fuentes O, Brunaugh AD, Kanapuram SR, Smyth HDC. Influence of formulation factors on the aerosol performance and stability of lysozyme powders: a systematic approach. AAPS PharmSciTech. 2018;19:2755–66. https://doi.org/10.1208/s12249-018-0980-9.
82. Graf M, Ziegler CE, Gregoritza M, Goepferich AM. Hydrogel microspheres evading alveolar macrophages for sustained pulmonary protein delivery. Int J Pharm. 2019;566:652–61. https://doi.org/10.1016/j.ijpharm.2019.06.019.
83. Wallace SJ, Li J, Nation RL, Prankerd RJ, Boyd BJ. Interaction of colistin and colistin methanesulfonate with liposomes: colloidal aspects and implications for formulation. J Pharm Sci. 2012;101:3347–59. https://doi.org/10.1002/jps.23203.
84. Falciani C, Zevolini F, Brunetti J, Riolo G, Gracia R, Marradi M, et al. Antimicrobial peptide-loaded nanoparticles as inhalation therapy for Pseudomonas aeruginosa infections. Int J Nanomedicine. 2020;15:1117–28. https://doi.org/10.2147/IJN.S218966.
85. Nieto-Orellana A, Coghlan D, Rothery M, Falcone FH, Bosquillon C, Childerhouse N, et al. Dry-powder formulations of non-covalent protein complexes with linear or miktoarm copolymers for pulmonary delivery. Int J Pharm. 2018;540:78–88. https://doi.org/10.1016/j.ijpharm.2018.02.008.
86. Tenland E, Pochert A, Krishnan N, Umashankar Rao K, Kalsum S, Braun K, et al. Effective delivery of the anti-mycobacterial peptide NZX in mesoporous silica nanoparticles. PLoS One. 2019;14:e0212858. https://doi.org/10.1371/journal.pone.0212858.
87. Garcia-Fernandez A, Sancenon F, Martinez-Manez R. Mesoporous silica nanoparticles for pulmonary drug delivery. Adv Drug Deliv Rev. 2021;177:113953. https://doi.org/10.1016/j.addr.2021.113953.

Pulmonary Delivery of Nucleic Acids

Gemma Conte, Ivana d'Angelo, Joschka Müller, Benjamin Winkeljann, Simone Carneiro, Olivia M. Merkel, and Francesca Ungaro

Abstract Pulmonary diseases affect millions of people worldwide and elicit symptoms that may evolve into severe conditions. Despite currently recommended treatments that can successfully control mild to moderate symptoms, targeted therapy has the potential to improve serious lung conditions over long time spans. Nucleic acid therapy has shown remarkable efficacy outcomes against pulmonary diseases through its mechanism of modulating intracellular gene expression. To improve the stability of nucleic acid molecules under physiological conditions and achieve targeted delivery, nucleic acids engineered in drug-delivery systems and administered by inhalation have been investigated in recent years. In this chapter, we will review the recent advances in the inhalation of nucleic acid-based therapies for lung disorders. The manufacturing of nucleic acid-based nanosystems for inhalation, as well as their potential to overcome physiological barriers, will be discussed. Ultimately, we will report on studies on innovative inhaled formulations that achieved promising results in animals or in clinical trials and summarize the promising strategies for pulmonary diseases.

G. Conte
Department of Pharmacy, University of Napoli Federico II, Naples, Italy

Di.S.T.A.Bi.F., University of Campania Luigi Vanvitelli, Caserta, Italy

I. d'Angelo
Di.S.T.A.Bi.F., University of Campania Luigi Vanvitelli, Caserta, Italy

J. Müller · S. Carneiro
Department of Pharmacy, Pharmaceutical Technology and Biopharmacy, Ludwig-Maximilians-Universität München, Munich, Germany

B. Winkeljann
Department of Pharmacy, Pharmaceutical Technology and Biopharmacy, Ludwig-Maximilians-Universität München, Munich, Germany

Center for NanoScience (CeNS), Ludwig-Maximilians-Universität München, Munich, Germany

J. K. W. Lam, P. C. L. Kwok (eds.), *Respiratory Delivery of Biologics, Nucleic Acids, and Vaccines*, AAPS Introductions in the Pharmaceutical Sciences 8,
https://doi.org/10.1007/978-3-031-47567-2_5

Keywords Inhalation · Nucleic acids · RNA therapeutics · Lung disorders · Nanoformulation

1 Introduction

Respiratory diseases are a global burden, with millions of deaths attributed to pulmonary illnesses and dysfunctions. The constant progress in the medical field has oriented attention toward innovative pharmacological therapies for severe lung diseases. Thus, conventional therapeutic approaches are currently taking a backseat in favor of therapeutic nucleic acids (NAs) as viable alternatives for the treatment of undruggable targets, with the purpose of acting at the intracellular or even nuclear level.

NAs are biological macromolecules, among which DNA and RNA separately regulate intracellular gene expression [57]. From a chemical point of view, NAs are negatively charged macromolecules due to the presence of a sugar-phosphate backbone composed of single or double strands characterized by 20 to 1000 bases each. Naked NAs have a half-life of less than an hour in human plasma and are often unable to penetrate cellular membranes [100]. They suffer nuclease-mediated degradation as well as inherent chemical instability, which are among the major bottlenecks in the progress of NA-based pharmaceutical product development.

Recent studies demonstrate the possibility to chemically-modify nucleosides to improve metabolic stability, therapeutic outcome, and reduce immunogenicity [18, 108]. In general, sugar, phosphate-linkage, and base modifications are commonly used [146]. Sugar-modified NAs with a 2′-*O*-methyl, 2′-*O*-methoxyethyl, locked nucleic acids (LNA) or unlocked nucleic acids (UNA) can efficiently decrease immunogenicity [134], improve specificity and activity, or reduce site-specific off-target effects [19, 50] Phosphate-linkage modifications mainly increase the stability toward nucleases, resulting in an increased circulation half-life [139]. Base modification, including the substitution of pseudouridines, 2-thiouridines, N6-methyladenosines, and 5-methylcytidines, results in reduced toxicity and enhanced biological stability, compared to unmodified RNA [9, 77].

O. M. Merkel (✉)
Department of Pharmacy, Pharmaceutical Technology and Biopharmacy, Ludwig-Maximilians-Universität München, Munich, Germany

Center for NanoScience (CeNS), Ludwig-Maximilians-Universität München, Munich, Germany

Institute of Lung Health and Immunity (LHI) and Comprehensive Pneumology Center (CPC) with the CPC-M bioArchive, Helmholtz Munich, Member of the German Center for Lung Research (DZL), Munich, Germany
e-mail: Olivia.merkel@lmu.de

F. Ungaro
Department of Pharmacy, University of Napoli Federico II, Naples, Italy

In parallel to NA chemical modification, it is paramount to develop safe and effective delivery systems that are able to carry molecules into the target cells in an efficient, safe, and reproducible manner. With this idea in mind, studies investigating local NA delivery for the treatment of pulmonary diseases have rapidly expanded. The inhalation route quickly turned into the first choice to allow direct delivery of NAs to the target site of action while minimizing systemic exposure. Advanced delivery strategies based on viral and nonviral vectors are currently being investigated to improve NAs transport to the target in the lungs [73, 158]. Although viral vectors have various advantages, their clinical applicability and approval by regulatory authorities are still hampered by safety (integration, immunogenicity), manufacturability (consistent large-scale production), and efficacy concerns [144]. On the other hand, the rapidly growing sophistication of inhalable nonviral carriers opens new opportunities aimed at improving NAs' pulmonary bioavailability [73].

This chapter will focus on inhaled NA-based therapies for pulmonary diseases with particular consideration on the hurdles imposed by the administration route. Moreover, the recent progress in engineering and producing nucleic acid-based nanoplatforms for inhalation will be discussed.

2 Nucleic Acid Therapeutics for Inhalation

Since the discovery of NAs, their therapeutic potential in multiple diseases has been extensively demonstrated. In recent years, NA-based therapies have become a very attractive therapeutic modality for various respiratory diseases [98]. Amid them, those having a genetic origin, such as cystic fibrosis (CF), alpha-1 antitrypsin deficiency (AATD), and primary ciliary dyskinesia (PCD), may strongly benefit from inhaled NAs. Nonetheless, NAs, and especially RNA therapeutics, are also very promising for the treatment of airway inflammation and hyperresponsiveness and have been suggested for therapy against chronic obstructive airway disease (COPD) and asthma [153]. These are also potential fibrotic lung diseases, which represent a large subset of severe lung diseases with unmet clinical needs, where NA-based therapies can make the difference, inhibiting pathways that are otherwise difficult to target [117]. This is the case of idiopathic pulmonary fibrosis (IPF), an aggressive type of interstitial lung disease with a very poor prognosis.

NAs of interest for the prevention and treatment of severe lung diseases can be broadly divided into two functional classes on the basis of the induced pathway (Fig. 1): (i) gene downregulation (or protein knockdown) pathway (e.g., antisense oligonucleotides, microRNA, small interfering RNA); (ii) protein expression pathway (e.g., messenger RNA). In recent years, the clustered regularly interspaced short palindromic repeats (CRISPR) and CRISPR-associated protein 9 (Cas9) genome editing system has also emerged as a versatile tool to achieve gene insertions/deletions, gene replacements, and single base pair conversions [28].

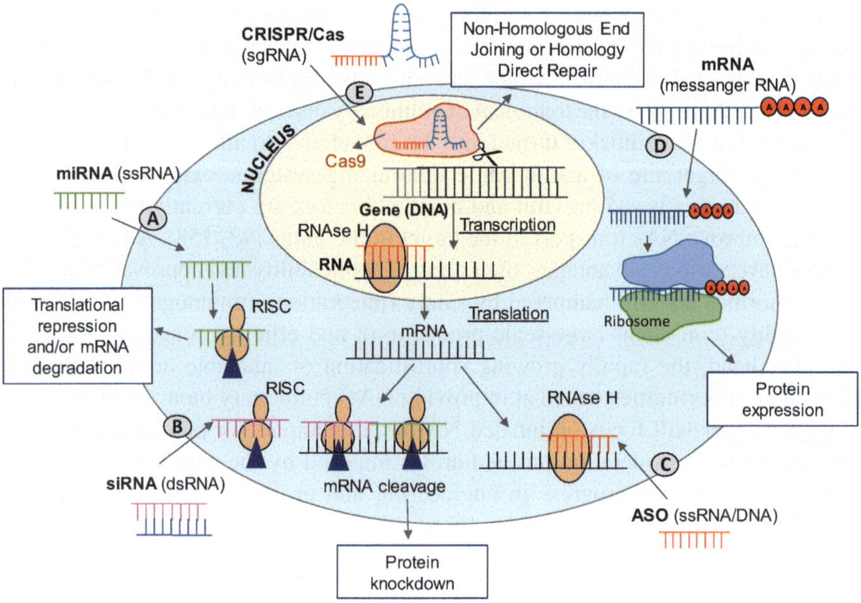

Fig. 1 Schematic representation of main NAs functional classes based on the induced pathway. (*A*) miRNA, single-stranded RNA (ssRNA) binding to RNA-induced silencing complexes (RISC) and inducing translational repression and/or mRNA cleavage. (*B*) siRNA (synthetic double-stranded RNA) acting on RNAi pathway inducing mRNA cleavage and consequently protein knockdown. (*C*) Antisense oligonucleotides (ssRNA or ssDNA) acting on cleavage of target mRNA through ribonuclease H (RNAse H) complex. (*D*) mRNA binding the ribosome is subsequently translated into proteins. (*E*) Genome editing mediated by single guide RNA (sgRNA) and Cas9 enzyme (CRISPR/Cas genome editing)

2.1 Antisense Oligonucleotide (ASO)

ASOs are single-stranded DNA or RNA molecules with sizes ranging from 13–25 bases. ASOs bind selected mRNA through the activation of the endonuclease ribonuclease H (RNAse H) [54], leading to degradation of the complementary mRNA and consequent downregulated translation of the target gene [110].

ASO hybridization specificity with target mRNA was increased by chemical modification at the 2′- hydroxyl group of the sugar [24], resulting in an altered RNAse H-mediated cleavage process [122]. Moreover, there are supplementary mechanisms associated with ASO activity, e.g. reducing telomeric activity or displacing mRNA pairing with specific transcription factors [13, 84].

Due to their promising activity, ASOs were the first NAs included in clinical trials, and nowadays are the most broadly approved NAs in therapy. Some naked (modified) ASOs were previously under evaluation for pulmonary disorders as in the case of TPI ASM8, SB010 and AIR645 from Pharmaxis (NCT00822861) [30], Sterna Biologicals Gmbh & Co (NCT01743768) [31] and Altair Therapeutics

(NCT00941577) [32]. All three drugs completed Phase 2 clinical trials for asthma, but several examples were suspended afterwards. TPI ASM8 is an inhaled ASO drug product, which attenuated both early and late asthma responses via inhibition of the target gene mRNA of chemokine receptor 3, and the common β chain of interleukin-3, interleukin-5 and granulocyte-macrophage colony-stimulating factor [52, 67]. Sterna's proprietary active pharmaceutical ingredient hgd40 is the active ingredient of SB010, a special type of catalytic 34mer ASO, a DNAzyme that specifically cleaves GATA-3 mRNA of humans and tox species [27].

2.2 *MicroRNA (miRNA)*

miRNA is a short noncoding single-stranded RNA (sizes range from 18–24 bases). Similar to siRNA, miRNA also acts via the intracellular RNAi pathway but differs with regards to nonspecificity of mRNA pairing [88]. The activation of the RNAi complex is induced by miRNA endonuclease Argonaute (AGO) activation, favoring the complementary mRNA cutting (as mentioned below for the siRNA mechanism) [58, 127]. miRNA directly binds the 3′-untranslated region (3′-UTR) of the mRNA but with partial complementary pairing. The ability to simultaneously downregulate a different target gene could be a potential therapeutic advantage due to the ability to eradicate a multiple target disease [12, 76].

Several studies reported various levels of miRNA expression in the lung, underlying their contribution to biological processes during organ development and pathogenesis of different lung diseases, such as COPD, asthma, pulmonary fibrosis, and pulmonary arterial hypertension [17]. As a consequence, miRNAs are receiving remarkable attention to be used as biomarkers in the clinical diagnosis of lung diseases. Meanwhile, miRNA-based therapeutics are being developed. For example, Remlarsen (MRG-201) from miRagen therapeutics is a synthetic miRNA mimicry of miRNA-29b, currently in Phase II clinical trial for fibrosis and idiopathic pulmonary fibrosis (NCT03601052) [33].

2.3 *Small Interfering RNA (SiRNA)*

siRNA is a short double-stranded RNA with sizes within the range of 21–23 base pairs. The intracellular pathway starts with the antisense strand binding to its complementary mRNA in the RNA interference (RNAi) system described above [6, 57]. RNAi is a highly conserved pathway, first described by Fire et al. in [49], found in a variety of eukaryotic cells [5, 49] – a breakthrough discovery, which was awarded the Nobel Prize in Physiology or Medicine in 2006. Fire et al. demonstrated for the first time that the expression of a target gene in the nematode *Caenorhabditis elegans* can be sequence-specifically inhibited by the introduction of double-stranded

RNA (dsRNA) [49]. In this process, a long double-stranded RNA sequence is recognized by the enzyme RNAse III Dicer and cut into smaller fragments [15]. Subsequently, these fragments unwind and bind to the endonuclease Argonaute type 2 (AGO2), building the RNA-induced silencing complex (RISC) [1, 84]. While siRNA enters the RISC as double-stranded RNA, AGO2 separates the complementary RNA strands and releases the passenger RNA into the cytosol. This activated complex can theoretically cleave any target RNA, which is complementary to the loaded siRNA, leading to specific RNA degradation [1, 2].

In 2018, the world's first RNAi-based drug was approved by the U.S. Food and Drug Administration (FDA). Alnylam Pharmaceutical's ONPATTRO® was granted a Breakthrough Therapy Designation (BTD), certifying an exceptional achievement. ONPATTRO® is used for the treatment of adults with hereditary transthyretin-mediated ATTR amyloidosis (hATTR amyloidosis) with polyneuropathy. At the time this chapter is being written, other Alnylam and Novartis siRNA therapeutics have been approved in the European Union and by the FDA (i.e., Givosiran for Acute hepatic porphyria, Lumasiran for Primary hyperoxaluria type 1, and Inclisiran for Familial Hypercholesterolemia). Most recently, Vutrisiran (AMVUTTRA®) was approved on June 13, 2022, for the treatment of the polyneuropathy of hATTR amyloidosis, indicating that research and development in this area is presently developing.

Based on these observations, RNAi can offer a therapeutic approach for diseases that currently cannot be treated with small molecules or therapeutic antibodies. One advantage is that RNAi uses a naturally occurring biological signaling pathway that is available in probably every cell in the human body [5]. In addition, any siRNA sequence can be chosen freely, which gives the possibility to knock out theoretically any gene. siRNA has a catalytic effect, i.e., one molecule can mediate the degradation of several RNA molecules [64]. Furthermore, RNAi drug candidates can be identified using bioinformatic tools that select sequences complementary to the target mRNA [64]. This in silico method of identifying suitable sequences is advantageous and resource-saving compared to conventional screenings being necessary for drug discovery of conventional drugs such as small-molecule inhibitors or antibodies.

siRNA-mediated gene silencing also bears hurdles to be overcome. Since free RNA may be degraded by exo- and endo-nucleases, siRNA stability is one of the most challenging issues [70]. Even with chemically modified bases, it cannot be excluded that the body responds with an innate immune response to the administered siRNA dose [45]. Another important point to consider is the suppression of off-target genes. In previous work, Jackson et al. concluded that a match of 11 nucleotides is sufficient for the knockdown of nontarget mRNAs [69]. Therefore, it is even more important to identify suitable, target-specific sequences with the lowest possible match to the human genome.

Despite the huge therapeutic potential, only two clinical trials administering aerosolized siRNA have been undertaken so far (ALN-RSV-01, Alnylam Pharmaceuticals, Phase IIb completed; Excellair™, ZaBeCor Pharmaceutical Co,

Phase II discontinued). In 2007, the Phase I clinical trial on ALN-RSV01 (NCT00496821) [34] was the first to investigate the potential of siRNA for inhalation and demonstrated its tolerability following intranasal administration. The Phase IIb clinical trial (NCT01065935) [35] further showed that aerosolized ALN-RSV01 was effective in lung transplant patients with respiratory syncytial virus infection following inhalation. Nonetheless, ALN-RSV01 failed to progress to a Phase III trial due to missing clinical endpoints. Similarly, a Phase I trial in 2009 demonstrated that Excellair™ was well-tolerated in patients with asthma. Nevertheless, the Phase II trial was discontinued in 2015. Only recently, a double-blind, randomized, placebo-controlled Phase I study was undertaken to assess the safety and tolerability of single and multiple inhaled doses of a siRNA against transforming growth factor-beta 1 (TGF-β1) (TRK-250, Toray Industries, Inc./BONAC Corporation) in subjects with IPF [36]. Despite representing important milestones of inhaled siRNA therapy, the failure of both products highlights the intrinsic difficulties involved with the translation of inhaled siRNA from the bench to the clinics.

2.4 Messenger RNA (mRNA)

Nowadays, the use of mRNA in treating lung diseases has gained particular attention, especially after the FDA approval of COVID-19 mRNA-based vaccines. mRNA is single-stranded RNA implicated in protein translation. More precisely, it is able to transfer protein information from the nucleus to the cytoplasm. Interfering with the mRNA pathway allows to predict the control of protein expression and to act on undruggable targets [128].

The main advantage of mRNA-based drugs in general is the ability to escape any kind of off-target effect, since its activity occurs directly at the cytosolic level and exploits the cells' own machinery. Nonetheless, the delivery of naked mRNA is heavily hampered by endonuclease-induced degradation. In this regard, numerous studies have been carried out to improve physiochemical stability in the lung environment [120].

In May 2018, MRT5005 from Translate Bio Inc., an mRNA encoding fully functional cystic fibrosis transmembrane conductance regulator (CFTR) protein, was the first of its kind to reach clinical testing. A Phase I/II clinical trial in patients with cystic fibrosis (CF) demonstrated its safety and tolerability, and interim results of therapeutic efficacy were encouraging as well. Nonetheless, a small clinical trial in early 2021 has found no pattern of improvements in lung function in CF patients who received MRT5005.

Allowing proper delivery of NAs to an undruggable target is of course paramount. Therefore, the need for drug-delivery systems able to improve NA stability, to shield its interaction with the lung environment and to increase the therapeutic effect becomes obvious [57, 115].

2.5 CRISPR/Cas

CRISPR-Cas9 is ribonucleoprotein (RNP) made of the protein Cas9 and a chimeric single-stranded RNA, which is referred to as "single guide RNA" (sgRNA). Active CRISPR/Cas enables the cells' pathways for genome repair, which can be either Non-Homologous End Joining (NHEJ) or Homology Direct Repair (HDR) [91, 93]. The Nobel Prize-awarded technology has revolutionized gene editing on the lab scale and is gaining interest in therapeutic approaches. Therapeutic delivery of both components, however, is challenging but can in principle be achieved in three different ways [46, 155], namely delivery of (i) Cas9 protein and sgRNA, (ii) plasmid DNA encoding for Cas9 protein and sgRNA, or (iii) Cas9 mRNA and sgRNA. Most in vivo studies focus on systemic administration of CRISPR/Cas, and accordingly, the first account in the literature reporting on a potential treatment of lung pathologies exploits systemic administration of LNPs delivering CRISPR-Cas9 with a passive accumulation in the lungs. Lung accumulation was reported by Parhiz and colleagues [114] and by using the selective organ targeting (SORT) technology, where increased cationic lipid content in the formulation redistributes LNPs to the lungs [26, 129]. Only recently, the first report on nebulized Cas13a delivery was published, where co-delivery of Cas-expressing mRNA with CRISPR RNA (crRNA) and trans-activating crRNA (trRNA) yielded therapeutic reduction of influenza and SARS-2 viral RNA. Unfortunately, the therapeutic outcome was analyzed only based on viral titers and body weight, while histology or lung function was not further assessed [16].

3 Overcoming Lung Barriers to Nucleic Acids for Inhalation

Pulmonary delivery offers several advantages over systemic administration, such as reduced first-pass effect, local administration with decreased systemic side effects, and noninvasive application. Nonetheless, the main lesson learned from the recent pandemic is that lungs have developed an efficient multi-tiered defense system to block any foreign substance from causing damage, including inhaled NAs. The effectiveness of any lung-targeted therapy is dramatically reduced by physical and biological innate barriers. After inhalation, a key role is played by macrophages and other cells of the immune system, which are involved in the defense mechanism against any foreign material that impacts the lung. Additionally, the noncellular components of the lung barrier (i.e., airway mucus, lung surfactant) are paramount to allow drug translocation across the lung epithelium and macrophage uptake. In fact, the behavior of inhaled particles in the lung-lining fluids determines their solubility, dissolution rate and/or diffusion toward their target cell, and hence their availability (Fig. 2).

The barriers imposed by the pulmonary route can be broadly divided into two main categories: (i) *noncellular barriers* and (ii) *cellular barriers* [59].

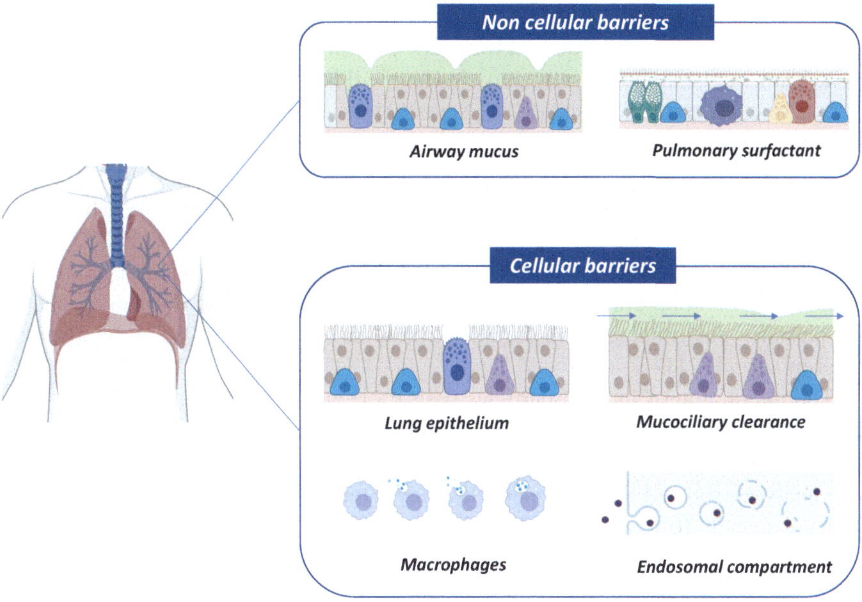

Fig. 2 Schematic representation of noncellular and cellular barriers in the lungs

3.1 Noncellular Barriers

The lung-lining fluids covering the respiratory epithelium are a primary and heterogeneous constituent of the pulmonary host defense mechanism. According to their location throughout the respiratory tract, two main elements of the lung-lining fluid have been identified: the airway surface liquid (ASL), a mucus gel-aqueous sol complex lining conductive airways, and the fluid lining the alveoli, a complex of alveolar subphase fluid (AVSF) and pulmonary surfactant [74].

Airway mucus is one of the most investigated noncellular barriers affecting the persistence and the extent of absorption of any inhaled drug in the lungs [38, 133]. It is a hydrogel composed mainly of water (95%), glycoproteins (wherein the main component is mucin), salts, non-mucin proteins, lipids, DNA, and cellular debris. Among the organic constituents of airway mucus, mucins are the most relevant macromolecules (e.g., MUC5AC and MUC5B), responsible for the formation of the interconnected molecular network that gives the mucus its typical viscoelastic properties [10]. Two major mechanisms prevent drug particles from readily diffusing through mucus gel, that are "size filtering" through the mucus meshes and "interaction filtering" [38, 133].

Insoluble particles can be filtered by the mucus gel layer if they are bigger than the mesh-spacing of the mucin network. Otherwise, they establish hydrophobic, electrostatic, and/or hydrogen bonding interactions with the negatively charged mucin chains. Trapped particles are moved toward the pharynx and ultimately to the gastrointestinal tract by the upward movement of mucus generated by the beating cilia (i.e., mucociliary clearance) [105]. The pore size of mucus gels is around 100 to 200 nm, suggesting that only particles below this size range can potentially diffuse across the mucus layer [138]. Nevertheless, significant healthy-to-diseased and/or patient-to-patient variations are reported [65, 96].

Following deposition into alveoli, inhaled particles interact with *pulmonary surfactant* (PS), a mixture of lipids (90%), and surfactant proteins (90%), which prevents alveolar collapse during expiration [121]. Upon contact with PS, larger particles are displaced from the airspace to the hypophase due to wetting forces, a phenomenon that probably also occurs with nanosized particles. In the hypophase, the particles may interact with surfactant proteins or may be taken up by alveolar macrophages [121].

The interaction of inhaled particles with PS is another emerging challenge [51, 94] and demands robust and reproducible in vitro models. A few animal-derived surrogates of PS are commercially available, and combined proteomic and lipidomic analyses showed that nanoparticles may interact with a range of PS proteins, with consequent formation of a "hard" corona, likely affecting interactions with the biological milieu [121]. Many literature findings also suggest that, depending on their composition and size, inhaled drugs may interfere with the function of pulmonary surfactant, thus hindering its physiological and essential role in the lung [67, 124].

3.2 Cellular Barriers

Inhaled drugs that successfully circumvent possible interactions with lung-lining fluids may then be absorbed into the lung tissue. This aspect appears crucial in NA-based therapies, where the drug target is located inside the cells. The human lung epithelium is an important regulator and effector tissue, which provides a physical barrier to inhaled drugs and participates in drug clearance processes through the mucociliary escalator and lung macrophages. Phagocytosis of delivered particles by resident macrophages induces widespread pulmonary and systemic inflammation, which can significantly limit any therapeutic effects of NA delivery. Furthermore, if cell targets of therapeutic NAs differ from epithelial cells, the epithelium itself becomes a major barrier, which is able to avoid the contact, and consequently, the transfection or transduction of subepithelial cells [11].

Lung Epithelial Barrier The human airway epithelial barrier is a pseudostratified epithelium of cells connected by tight junctions that exhibit different cellular types: columnar-shaped and ciliated cells, with epithelial type I (ET-1) and II (ET-II) cells in the alveolar tract. This barrier separates the lumen airspace from the pulmonary aqueous interstitial compartment, which contains a variety of cells, collagen, elastic fibers, interstitial fluid, and lymphatic vessels. The mechanism of drug uptake by lung epithelial cells after deposition has not yet been fully understood [126]. While lipophilic drugs are thought to be rapidly absorbed by passive transcellular diffusion through epithelial cells, small hydrophilic compounds likely diffuse across the epithelium through aqueous pores in intercellular gap junctions. Lung physiology and lung pathology [150] clearly influence the extent of drug absorption, as in case of tight junctions, efflux proteins and cellular enzymes that play an important role as barriers in the absorption process [20].

To improve drug accumulation at the cellular level, different formulation approaches can be envisaged. Of course, carriers with adequate size and surface properties (charge, hydrophilicity, shielding cloud) may regulate drug interactions with cells [101]. Nonetheless, efficient drug delivery to the final intracellular target is still challenging.

Macrophages and Mucociliary Clearance Mucociliary clearance (MCC) and lung macrophages have the function of eliminating any foreign material that reaches the lung environment [63]. MCC is the dominant drug removal mechanism in the upper airways [126]. This process is the result of a coordination between mucus, serous cells, secretory gland, and ciliated cells. Ciliated cells, the main component of the upper pulmonary routes, provide, together with cilia coordinated beating, the necessary force to induce MCC [22]. The efficiency of MCC depends on three main factors: the frequency of the beat, the coordination of the cilia, the amount and rheological properties of airway secretions (derived from surface and submucosal calyx cells glands).

Additionally, in the alveolar space, macrophages can strongly reduce the in-situ drug availability. Alveolar macrophages (AM) represent the lung immunological barrier [107] and are the predominant phagocytic and antigen-presenting cells in the human respiratory tract. AM clearance is dominant in the peripheral lungs and is associated with MCC [59, 126]. The mechanism and the efficiency of drug uptake by macrophages change as a function of the inhaled particle size, shape and stiffness. Although is predominately considered a side effect, the AM uptake of inhaled drugs or genes can be considered the final goal in macrophages-target therapies [102].

3.3 Endosomal Escape: Intracellular Level

Once NAs cross the cell membrane, the endosomal compartment comprises yet another hurdle the drug or the drug-delivery system must pass [130]. The endosomal-lysosomal system can be considered a sorting machinery within the cell. In brief, fluids, solutes or drugs or drug-delivery systems that are internalized by the cell are first trafficked to the early or sorting endosome. From there, the cargo might be sent to the recycling pathway to be exocytosed again. Alternatively, further trafficking over the multivesicular body (often referred to as late endosome) and endo-lysosome to the degrading environment of the lysosome comprises one of the standard routes.

The trafficking route within this complex compartment is thereby determined by a multitude of signaling molecules [62, 75, 132].

To be able to evade recycling or degradation of NA drugs, drug delivery scientists employ different physical and chemical mechanisms; yet most of these strategies aim at destabilizing or disrupting the endosomal membrane one way or another [21, 44, 60, 131, 141] (Fig. 3). More complex strategies aim at interfering with the signaling pathway in order to redirect the trafficking route, for example, to directly enter the nucleus [123]. The topic was recently reviewed in detail for nucleic acids therapeutics, and the reader is therefore referred to [148].

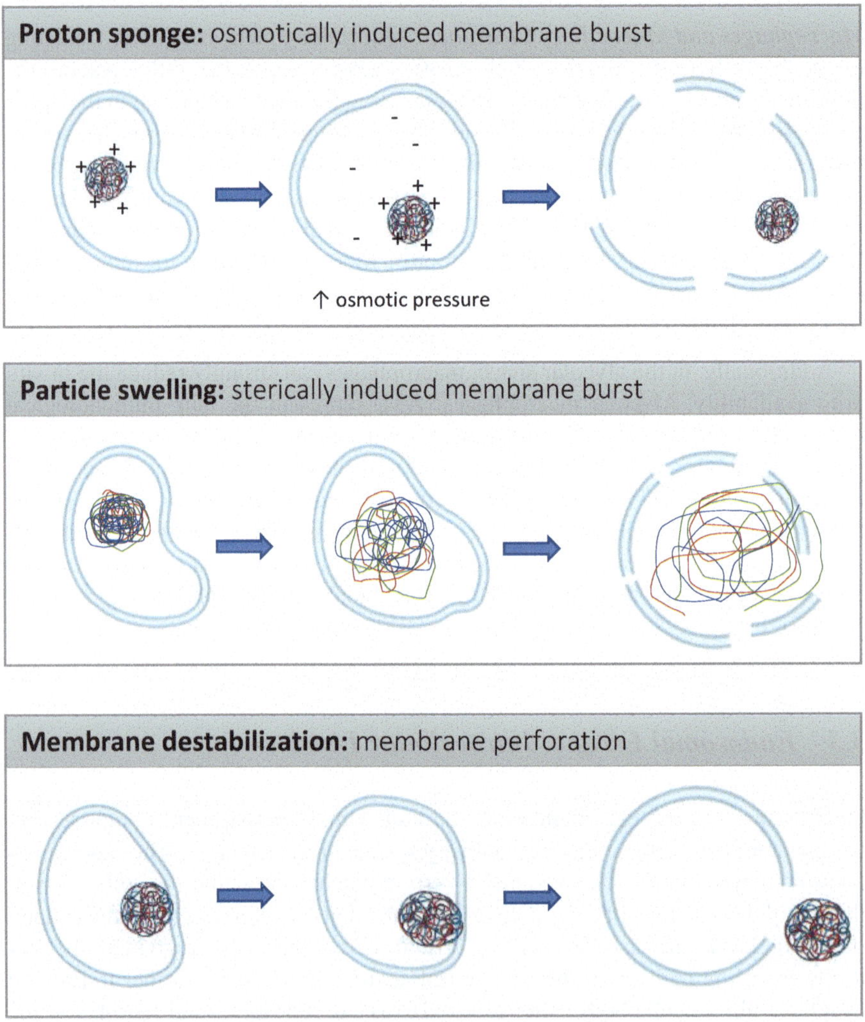

Fig. 3 Schematic representation of hypothesized endosomal escape mechanisms

The success rate of endosomal escape, however, remains low: studies found that in most cases, only a few percent of nanoparticles that accomplished being taken up by the target cell are, in the end, able to escape the endosome [53, 60, 135, 143].

4 Engineered Nanoplatforms for Nucleic Acid Inhalation: The Case of RNA-Based Therapeutics

Amid NAs, RNA-based therapeutics are receiving special attention for their diverse roles and potential therapeutic capacity. With the recent success of siRNA-based therapeutics in the clinic and the intensive investigation of mRNA in clinical trials, prompted by the approval of patisiran (ONPATTRO®) and mRNA vaccines against coronavirus disease 2019 (COVID-19), these two types of RNA are likely to be the first to enter the clinic for treating lung diseases.

Several nonviral vectors are being developed to optimize pharmacokinetics and biodistribution of inhaled RNAs, spanning from nanocomplexes (lipoplexes, polyplexes) to lipid-, polymer- or hybrid lipid/polymer nanoparticulate systems (Table 1). The ability to deliver NA cargo into airway epithelial cells is considered an important turning point for emerging inhaled RNA therapies.

4.1 Lipid-Based Nanoplatforms

Among different types of delivery systems, lipid-based nanoplatforms, and especially lipid nanoparticles (LNPs), have been extensively studied for RNA delivery due to their unique properties, such as simple synthesis of the lipid components, scalable manufacturing processes, and wide packaging capability [25, 144]. Whereas lipoplexes based on cationic lipids were popular as transfection agents during the early years of gene therapy, the first clinically effective NA nanoplatform was an LNP (i.e., ONPATTRO®).

Over the past decades, cationic liposomes have been the gold standard for RNA delivery, taking advantage of permanent positive charge to complex negatively charged NA, thus obtaining so-called "lipoplexes" [40]. This self-assembly is commonly achieved by the use of cationic lipids comprising a quaternary ammonium function, such as *N*-[1-(2,3-dioleyloxy)propyl]-n,n,n-trimethylammonium chloride (DOTMA) and 1,2-dioleoyl-3-trimethylammonium-propane chloride salt (DOTAP). Cationic lipids are often used in combination with helper lipids such as 1,2-dioleoyl-sn-glycero-3-phosphoethanolamine (DOPE) and cholesterol. Several parameters can affect the stability of this complex. For example, the lipid composition is a critical parameter to achieve protection of the RNA against body fluids and to mediate interaction with cell membranes, followed by endosomal escape, and finally the release of the RNA cargo in the cell cytoplasm [14, 40]. Despite advances in

Table 1 Representative inhaled nanoplatforms developed for NA delivery to the lungs and main in vitro/in vivo findings

Formulation	NA	Model	Main findings	Reference
Lipid-based nanoplatforms				
Lipoplexes	CFTR plasmid (pGM169)	Human studies (phase 2b clinical trial)	Lung function stabilized in some individuals; no significant adverse effects	Alton et al. [7]
Liposomes (receptor-targeted nanocomplex)	siRNA targeting αENaC	Female C57Bl6 mice	Effective and prolonged ENaC silencing in vivo	Tagalakis et al. [136]
PEGylated LNPs	Luciferase reporter mRNA (FLuc mRNA)	Balb/c mice	Enhanced penetration across airway mucus/cell barriers; no differences of in vivo luminescence intensity upon treatment with prenebulized or nebulized LNPs (i.e., resistance to shear stresses upon nebulization)	Zhang et al. [157]
LNP (modified with DMG-PEG β-sitosterol)	mRNA encoding CFTR	CFTR-deficient mice	Improved CFTR expression without pulmonary or systemic toxicity	Kim et al. [82]
Spray dried LNPs	siRNA targeting GAPDH	Ex vivo human precision-cut lung slices	Up to 50% gene silencing without any signs of toxicity	Zimmermann et al. [159]
Polymer-based nanoplatforms				
Spray dried PEI/DNA polyplexes (Threalose-based NEM)	GFP plasmid	A549 cells	Efficient uptake and transfection profiles upon NEM re-dispersion	Keil et al. [80, 81]
Self-assembled peptide-poloxamine complexes	SBTS consisting of SB transposon (pDNA) and transposase-encoding SB100X-mRNA	CFBE-delF cells CF mice	Long-term restoration of CFTR in CFBE-delF cells and CF mice	Guan et al. [55]
hPBAE/mRNA polyplexes	IVT mRNA encoding for firefly luciferase	Ai14 reporter mouse model	Consistent protein production in the lung, without local or systemic toxicity	Patel et al. [115]
	CRISPR RNAs	Influenza A - infected mice SARS-CoV-2-infected hamsters	Degradation of influenza RNA in mouse lung tissue reduction of SARS-CoV-2 replication and symptoms in hamsters	Blanchard et al. [16]

(continued)

Table 1 (continued)

Formulation	NA	Model	Main findings	Reference
PBATE	Different RNA cargos	Different animal models (mice, hamsters, ferrets, cows, rhesus macaques)	P76 delivers cargo of any size to the lungs of mice with minimal toxicity P76 delivers mRNA across species with minimal toxicity	Rotolo et al. [125]
Hybrid lipid/polymer nanoplatforms				
DOTAP/PLGA NPs	miRNA mimics or dsDNA inhibiting IL-8	Human NuLi-1 bronchial epithelial cells (BECs)	Effective and nontoxic carriers for nebulized delivery of miR-17 to BECs	Vencken et al. [142]
DPPC/PLGA NPs	siRNA pool targeting αENaC and βENaC	Triple cell co-culture Model (TCCC) grown at ALI and A549	Efficient internalization in TCCC model and prolonged and effective knockdown of both ENaC a and b subunits in A549	d'Angelo et al. [42]
DPPC/PLGA NPs and DSPE-PEG/ PLGA NPs	siRNA pool against NFkB	LPS-stimulated 16HBE14^{o-} Cells	PEGylation does not make a difference when the mucus barrier properties are dominated by pathology-associated proteins. Increased in vitro inhibition of NFκB gene overexpression in non-PEGylated NPs	Conte et al. [39]

Abbreviations: αENaC = α subunit of the sodium transepithelial channel; βENaC = β subunit of the sodium transepithelial channel; CFTR = cystic fibrosis transmembrane conductance regulator; DMG-PEG = 1,2-dimyristoyl-rac-glycero-3-methoxypolyethylene glycol; DOTAP = Dioleoyl-3-trimethylammonium propane; DPPC = Dipalmitoylphosphatidylcholine; DSPE-PEG = 1, 2-Distearoyl-sn-glycero-3-phosphoethanolamine-Poly(ethylene glycol); GAPDH = Glyceraldehyde-3-Phosphate Dehydrogenase; GFP = Green Fluorescent Protein; IVT = In vitro transcribed; LNP = lipid nanoparticles; LPS = Lipopolysaccharide; NEM = nano-embedded microparticles; NFκB = Nuclear Factor kB; PBAE = poly-β-amino-ester; PBATE = poly-β-amino-thio-ester; PEG = Polyethylene glycol; PEI = Polyethylenimine; SB = Sleeping Beauty; SBTS = Sleeping Beauty Transposon System

lipoplex formulation, clinical trials are still limited due to concerns with their toxicity and immunogenicity. To date, the most ambitious clinical trial involving cationic lipids for inhalation was initiated by Alton and colleagues in 2008. CFTR plasmid (pGM169) was formulated with Genzyme lipid 67 (GL67). The incorporation of small amounts of DMPE-PEG5000 enabled the preparation of lipoplexes with an optimal cationic lipid:pDNA ratio of 0.75:1 for aerosolization. Patients (n = 78) received the nebulized lipoplex once *per* month for 1 year. Lung function was modestly stabilized in some individuals, and no significant adverse effects were observed.

However, despite these encouraging results, the approach was not sufficient to achieve a clear phenotypic correction [7].

In the attempt to overcome some of the limitations related to conventional lipoplexes, innovative lipoplex formulations, comprising pH-sensitive lipids, peptides and cell-penetrating peptides (CPP) or polymers, are being investigated for RNA delivery [14]. Recently, cationic receptor-targeted nanocomplex (RTN) formulations comprising cationic liposomes (DOTMA:DOPE at 1:1 molar ratio), a targeting peptide moiety and a siRNA targeting αENaC were tested in murine models. RTN translocated more rapidly than siRNA alone through mucus. Transfections of primary CF epithelial cells with nanocomplexes reduced αENaC and βENaC mRNA by 30%. A single dose of siRNA silenced ENaC by approximately 30% in the mouse lung, which persisted for at least 7 days, while three doses of siRNA increased silencing to approximately 50% [136]. Clinical practice would benefit from safer nonviral vectors, avoiding patients' immune responses and toxicity, which is of major concern.

LNPs comprising cationic lipids, cholesterol, and polyethylene glycol (PEG)-modified lipids have been developed to improve the structural stability in comparison to lipoplexes/liposome delivery systems [25, 144]. If adequately engineered, LNPs may efficiently entrap RNA molecules and allow to overcome one major issue associated with lipid-based systems for pulmonary delivery, that is their poor structural stability and consequent disassembling in the lung environment [4]. First in vivo proof of concept of the potential of LNPs for mRNA delivery to the lungs was given by Pardi et al. [113]. Nowadays, increasing knowledge on lipids and lipid derivatives affords for a rationale design of LNPs for inhalation [157]. The introduction and the type of PEGylated lipids have been shown crucial to increase the shear resistance of LNPs during nebulization and to enhance their penetration across the mucus and cell barriers within the lungs [157]. In a comparative study, 1,2-distearo yl-sn-glycero-3-phosphoethanolamine-*N*-methoxy(polyethylene glycol) (DSPE-PEG) was found to negatively affect the stability of LNPs as a significantly higher aggregation level appeared after nebulization compared to formulations with 1,2-dimyristoyl-rac-glycero-3-methoxy(polyethylene glycol) (DMG-PEG) and 1,2-dimyristoyl-sn-glycero-3-phosphoethanolamine-N-methoxy(polyethylene glycol) (DMPE-PEG) [156]. While DMG-PEG allowed to enhance the superficial stability of LNPs, β-sitosterol provided LNPs with a polyhedral shape, facilitating endosomal escape. Thus, mRNA encoding the cystic fibrosis transmembrane conductance regulator (CFTR) was delivered after nebulization to a CFTR-deficient mice, resulting in pulmonary expression of this otherwise endogenous protein [82]. In another study, Tam et al. [137] highlighted how the composition of helper lipids in LNPs crucially modulates transfection efficiencies in airway epithelia, and in the murine respiratory system. In vivo studies showed that intranasal delivery of LNPs containing helper lipids 1,2-distearoyl-sn-glycero-3-phosphocholine (DSPC), 1,2-dioleoyl-sn-glycero-3-phosphocholine (DOPC), egg sphingomyelin (ESM) and 1,2-dioleoyl-sn-glycero-3-phospho-L-serine (DOPS) with luciferase mRNA resulted in a significant increase in luminescence expression in the nasal cavity and lungs. Similarly, an in vivo cluster-based iterative screening approach was used to identify LNP chemical characteristics that promote lung delivery by nebulization [95]. The authors concluded that (1) PEG is required for LNP formulation and (2) the combination of PEG-lipid molarity and helper lipid structure and charge influences delivery. Of note, nebulized delivery of an mRNA encoding a broadly neutralizing antibody targeting haemagglutinin via the optimized LNPs protected mice from a lethal challenge of the H1N1 subtype of influenza A virus and delivered mRNA more efficiently than LNPs previously optimized for systemic delivery.

Though in most cases, lipid-based nanoplatforms for pulmonary delivery are formulated to be delivered through nebulization, some dry powder for inhalation has also been successfully developed [106, 147]. The feasibility of engineering spray-dried LNP-based powders for siRNA delivery was recently demonstrated. Optimized spray-dried LNPs penetrated the lung mucus layer and maintained bioactivity, resulting in >90% protein downregulation with a confirmed safety profile in a lung adenocarcinoma cell line. Additionally, the spray-dried LNPs successfully achieved up to 50% gene silencing of the housekeeping gene GAPDH in ex vivo human precision-cut lung slices without any signs of toxicity as determined by cytokine levels [159].

4.2 Polymer-Based Nanoplatforms

Although dating back to the 1990s, the concept of using polymer particles for pulmonary delivery has evolved over time and is experiencing growing research interest in recent years [89]. A fundamental feature of polymer systems, which is only partly shared by lipid-based carriers, relies on their ability to exert a prolonged drug release. This is crucial to reduce the number of administrations and to increase patient adherence to complex therapeutic regimens required by chronic lung diseases. Furthermore, adequately designed polymer-carriers may allow to overcome various systemic and cellular barriers imposed on NA, including nuclease-mediated degradation, cell membrane, endosomal compartment, and nuclear membrane [86].

Polymeric nanoparticles represent a well-established platform for the encapsulation and delivery of a multitude of therapeutic molecules, including NAs, due to the versatility of polymer physiochemical properties as well as the variety of available production techniques, which can be selected in view of the specific drug cargo and intended application[48]. To be successful, design of inhalable NPs should take into consideration well-defined rules to overcome limitations associated with the pulmonary route of administration, providing adequate composition, size, and surface modification along with respirability.

Similarly to the lipid-based nanoplatforms introduced above, many polymeric NA carriers establish electrostatic interactions to complex their payload. One typically differentiates between polyelectrolyte complex (polyplex) and polyplex micelle (micelleplex) formation. The latter typically include hydrophobic moieties to stabilize the polyplexes whose structural integrity would otherwise solely depend on electrostatic interactions [47].

These electrostatic interactions, which are prerequisite for the formulation of both poly- and micelleplexes are in general established via protonable amine groups. Consequently, simple polyamines such as polyethyleneimine (PEI), polylysine (PLL), or poly-(amidoamine) PAMAM are among the most intensively studied polymer materials for NA delivery [40]. PEI-based polyplexes for example have shown great potential to be applied in the lungs, and the ease of modification allows for attaching shielding agents, targeting moieties or lytic peptides to further increase their efficacy [56, 78, 79].

Yet, due to the intrinsic toxicity that polyamines often inherit, it became clear early on that alternatives are necessary to establish polyplex systems as safe and efficient nanocarriers; poly(beta-amino ester) (PBAEs), for example became a popular alternative due to their favorable toxicity profile. Furthermore, poloxamines have been demonstrated to successfully deliver both RNA and DNA cargos for cystic fibrosis treatment into the lungs [55]. Different studies have shown impressive results with inhaled PBAE formulations [115], including the ability to deliver mRNA encoding Cas13a for mitigating influenza virus A and severe acute respiratory syndrome coronavirus 2 (SARS-CoV-2) infection in mice and hamsters [16]. In the attempt to improve the performance of PBAE nanoformulations, Rotolo et al. recently developed a poly-β-amino-thio-ester (PBATE) enabling efficient lung delivery of mRNA—regardless of cargo size and complexity—to mice, hamsters, ferrets, cows, and rhesus macaques [125]. P76 was found to be safe and well-tolerated and effective in the Syrian hamster model of SARS-CoV-2 infection.

In analogy to the delivery of other carrier systems, nebulization is the most commonly employed approach but also spray dried formulations are being investigated to directly deliver the cargo to the lung [80, 81]. Yet, detailed analyses of the influence of the aerosolization process on the integrity of the nanocarriers, pharmacokinetics of the inhaled formulations, and long-term safety of inhaled RNA therapy formulations are scarce [29].

4.3 Hybrid Lipid/Polymer Nanoplatforms

An early exploited strategy to achieve mucus-penetrating particles able to improve cellular uptake is represented by hybrid lipid-polymer NPs [39, 42, 142]. Hybrid NPs (hNPs) consist of a biocompatible/biodegradable core (e.g., PLGA core) and a lipid shell on the surface. To date, different lipids and production techniques have been tested regarding the fabrication of hNPs for the delivery of different drugs, with the aim of defining the right preparation method to achieve core-shell systems with optimized technological features (morphology, size, polydispersity, encapsulation efficiency, release kinetics) [103].

In case of siRNA encapsulation, a special emphasis was given to hNPs engineered with biocompatible phosphatidylcholine (PC), dipalmitoyl phosphatidylcholine (DPPC), DSPE-PEG or cationic DOTAP [37]. Additionally, in vitro/in vivo safety, immunogenicity, gene silencing effects of hNPs comprising lipid derivatives (i.e., lipidoids) have been addressed [43, 90]. Recently, it was demonstrated how the DPPC layer surrounding the PLGA core confers muco-inertia to hNPs, thus assisting the transport of the nucleic acid cargo across the mucus-covered human airway epithelial barrier [39, 42]. Nonetheless, the use of DSPE-PEG2000 (i.e., surface PEGylation) did not boost mucus penetration in complex and sticky mucus, such as cystic fibrosis sputa from patients with polymicrobial colonization [39]. Previous results pave the way to a new generation of mucus-penetrating nanoparticles, comprising a phospholipid shell rather than the conventional PEG shell, which is in the limelight of modern nanotechnology to overcome the lung mucus barrier [61]. It should be noted that a systematic study of the impact of the phospholipid layer composition on mucus- and cell-penetration of siRNA-loaded hNPs and their gene silencing efficacy is missing in the current literature. In order to improve overall NA uptake after hNP treatment, NAs precondensed with cationic polymers are often encapsulated into the core-shell structures. As discussed above, PEI can electrostatically form polyplexes with NA, improving the encapsulation and release from polymeric NPs and facilitating cellular internalization [85].

4.4 *Bioinspired Nanoplatforms*

To circumvent limitations of synthetic nanocarriers, room-temperature-stable inhalable lung-derived extracellular vesicles or exosomes (Lung-Exos) as mRNA and protein drug carriers are a formidable alternative [118]. Compared with standard synthetic nanoparticle liposomes (Lipos), Lung-Exos exhibited superior distribution to the bronchioles and lung parenchyma and were successfully delivered to the lungs of rodents and nonhuman primates by dry powder inhalation. In a vaccine application, severe acute respiratory coronavirus 2 (SARS-CoV-2) spike (S) protein-encoding mRNA-loaded Lung-Exos (S-Exos) elicited greater immunoglobulin G (IgG) and secretory IgA (SIgA) responses than its loaded liposome (S-Lipo) counterpart [119]. Importantly, S-Exos remained functional at room-temperature storage for one month. These results suggest that extracellular vesicles can serve as an inhaled mRNA drug-delivery system that is superior to synthetic liposomes.

4.5 *Inhaled mRNA Nanovaccines*

Over the past 2 years, mRNA vaccines have attracted increased scientific interest due to their crucial and successful role throughout the COVID-19 pandemic. Although high safety and efficacy profiles were demonstrated after billions of administered doses worldwide, the protection induced by these vaccines decreased over a few months after the vaccination [92]. Indeed, the intramuscular (IM) administration of vaccines elicits both humoral and cell-mediated immune responses but lacks arousing efficient mucosal immunity [149]. According to the literature, vaccines administered by IM injection for respiratory infections do not present a sterilizing effect. Therefore, the virus may be stored in the nasal cavity and easily spread among other patients, contributing to reinfection and persistent high infection rates [8]. For comparison, the subcutaneous (SC) environment is poorly vascularized and may lead to slow antigen processing that can limit vaccine efficacy and, thus, is not adequate for vaccines against respiratory infections [87]. To overcome these drawbacks and improve vaccine efficacy over time, intranasal (IN) administration is a promising approach and has shown encouraging results regarding overall efficacy. IN has some advantages such as avoiding needles that consequently eliminate the need for training healthcare professionals and the risk of injuries and pain. It should be noted that the most remarkable outcome of IN administration is the resulting local mucosal immunity as well as a long-lasting systemic immune response [8, 71].

The respiratory tract has an arsenal of features that benefit a high-efficacy vaccine response. The large surface pulmonary area is highly vascularized and presents a significant density of antigen-presenting cells (APC), such as dendritic cells, B cells, and alveolar macrophages [72]. Therefore, following IN vaccination, the target cells from the upper and lower respiratory tract are triggered to develop a broadly protective immune response. Consequently, high levels of immunoglobulin proteins, such as neutralizing IgG, local mucosal IgA, and T cells are detected. A study conducted by [145] [145], confirmed this pattern upon the administration of an inhalable vaccine against COVID-19 in mice. The main findings postulated that the vaccine allowed the detection of specific IgG antibodies and elicited a remarkable mucosal IgA response as well as $CD4^+$ and $CD8^+$ T cells [145]. Mainly due to the production of IgA, a local immune response is developed in the nasal cavity that may induce a sterilizing effect and contribute to blocking new infections [87]. Importantly, IM administration of mRNA vaccines against COVID demonstrated systemic peak levels of IgA, which were lower and decreased faster than IgG serum levels. In contrast, IN vaccines tend to produce high local IgA levels that will remain longer in the mucosa, thus, conferring a mucosal immunity that is required for a sterilizing effect [71].

To ensure such a potent immune response, vaccines for IN administration require a design of formulation that will proportionate an increased penetration within the lungs and an efficient endosomal escape. In this context, LNPs are highlighted as the most advanced clinical platform available. Indeed, the two mRNA-based vaccines against COVID-19 currently approved take advantage of LNPs as a biocompatible and efficient vehicle. As already pointed out above, LNPs can efficiently encapsulate mRNA to ensure its protection against degradation, can penetrate the multiple barriers within the lungs, can easily interact with cells to be taken up, and are well-tolerated by patients after administration [154]. Meanwhile, other nanovehicle platforms have been investigated with encouraging preclinical outcomes upon IN administration [87, 145]. Regardless of the nanocarrier chosen for an inhalable vaccine, the physicochemical characteristics must be controlled during the manufacturing process to ensure optimal performance.

5 Development of Dosage Forms for Lung Delivery (Nebulization, pMDIs, DPIs)

The mechanical method for administering RNAs by inhalation is another important challenge for developing inhaled RNA therapeutics. In order to produce an inhalable pharmaceutical product, most currently available devices can be divided into three categories: pressurized metered dose inhalers (pMDIs), dry powder inhalers (DPIs), soft mist inhalers (SMIs), and nebulizers, which can be further classified into jet nebulizers, ultrasonic nebulizers and vibrating mesh nebulizers (VMNs) [41, 99]. Jet nebulizers aerosolize an aqueous solution by passing compressed air through a "venturi" nozzle. The accelerated air then encounters the bulk drug-containing liquid, whereby the high velocity of the air compared to the solution causes droplet formation. The primary droplets thus obtained are usually very large. Droplets that are too large are recirculated and returned to the reservoir, after which they are nebulized again by the compressed air. The recirculation, however, exposes the solution to shear stress, which can make jet nebulizers unsuitable for formulations prone to degradation [97]. It was shown that in many cases, 99% of the aerosol has to be recycled, which emphasizes that repeated shear forces affect the contained drug heavily [3]. Ultrasonic nebulizers use a piezoelectric crystal to generate acoustic waves that create oscillatory pressure differentials in the solution. The bulk liquid is aerosolized in relatively large droplets [23]. In case of polyplexes, VMNs have been used, which exert less shear stress on the nanocarrier than the other nebulizers [109]. In VMNs, a thin perforated membrane, called mesh, is vibrated by a piezoelectric crystal. The holes of the membrane are tapered, with the larger opening on the side of the liquid. Through the vibration, the liquid in the pores is ejected, which forms the aerosol [83].

For optimal drug inhalation, the nanoplatforms should be exposed to as little shear stress as possible. Studies have shown that VMNs in particular represent the mildest conditions [83]. Though lower shear stress can exert a positive effect on particle integrity and transfection efficiency, the stability of nanocarriers during nebulization is still an open issue, especially for lipid-based nanoplatforms, demanding for further and deeper investigations. In view of the existing challenges related to the use of nebulizers, attempts have also been made to produce DPIs for NA inhalation. As described above, most of the inhaled RNA therapeutics being developed are nanosized particles, which are commonly exhaled by inhalation. To overcome this issue, dry powders for inhalation can be envisaged by embedding nanocarriers into microparticles based on inert carrier by spray drying, referred to as "nano-embedded microparticles" (NEM) or "nano-in-microparticle dry powders" [116, 152, 159].

Various particle engineering methods and excipients (e.g., mannitol, leucine, lactose) can be used to optimize the physicochemical properties of NEM, including moisture content, density, particle size, surface roughness, particle shape, and solid-state properties. These features inevitably affect dry powder aerosol performance, which, along with powder stability, are especially critical to ensure the delivery of a reproducible dose to the airways [153]. Some examples of these approaches are reported in Table 2.

Table 2 Summary of main dry powder formulations for NA inhalation

Formulation architecture	Encapsulated molecule	Inert carrier	Production technique	Reference
Dry powder containing pDNA liposomes or lipoplexes	pDNA	Bovine serum albumin, leucine, mannitol	Spray-freeze drying	Tsukamoto et al. [140]
Dry powder containing pDNA or siRNA/chitosan polyplexes	pDNA or siRNA	Mannitol	Supercritical fluids	Okamoto et al. [111] and Ihara et al. [66]
Dry powder containing siRNA/PEI polyplexes	siRNA	Leucine, mannitol	Spray-freeze drying	Okuda et al. [112]
Dry powder containing naked siRNA	siRNA	Mannitol	Spray drying	Wu et al. [151]
Dry powder containing DOTAP modified poly(glycerol adipate-co-ω-pentadecalactone) nanoparticles	miRNA	Leucine, mannitol	Spray drying	Mohamed et al. [101]
Dry powder containing pDNA/PEI polyplexes	pDNA	Hyaluronic acid, leucine, mannitol, phenylalanine	Spray-freeze drying	Ito et al. [68]
Dry powder containing LNPs	siRNA	Lactose	Spray drying	Zimmermann et al. [159]
Dry powder containing peptide/pDNA complexes	pDNA	Mannitol	Spray drying	Munir et al. [104]

6 Summary and Outlook

Despite extensive preclinical studies and success, the number of clinically approved DNA/RNA therapeutics is still very low, and the number of failing clinical trials is comparably high. Sometimes, this has raised the question of whether the gene therapy field is worth investing more. Nonetheless, the main lesson learned from the recent COVID-19 vaccine development is that nanotechnology can make a difference in NA delivery. Meanwhile, lung delivery may offer a strategy for noninvasive administration of the cargo directly at the target site. Thus, it is important to identify the challenges inhaled nanomedicine is facing and to have a clearer view of how to tackle them. Though the development of inhaled nanoparticles has come a long way over the last decades, many aspects still need to be elucidated to boost research in this field. In perspective, a lead candidate RNA nanoparticle could be moved into the next phase of the drug development process very quickly as seen with the COVID-19 vaccine development.

Acknowledgments JM, BW, and OMM acknowledge funding from Volkswagen Foundation under contract number AZ-9A872. SC is a Georg-Forster Fellow and acknowledges the Alexander von Humboldt Foundation.

References

1. Aagaard L, Rossi JJ. RNAi therapeutics: principles, prospects and challenges. Adv Drug Deliv Rev. 2007;59:75–86.
2. Agrawal N, Dasaradhi PVN, Mohmmed A, Malhotra P, Bhatnagar RK, Mukherjee SK. RNA interference: biology, mechanism, and applications. Microbiol Mol Biol Rev. 2003;67:657–85.
3. Agu RU, Ugwoke MI, Armand M, Kinget R, Verbeke N. The lung as a route for systemic delivery of therapeutic proteins and peptides. Respir Res. 2001;2:198–209.
4. Albertsen CH, Kulkarni J, Witzigmann D, Lind M, Petersson K, Simonsen JB. The role of lipid components in lipid nanoparticles for vaccines and gene therapy. Adv Drug Deliv Rev. 2022:114416.
5. Almeida R, Allshire RC. RNA silencing and genome regulation. Trends Cell Biol. 2005;15:251–8.
6. Alshaer W, Zureigat H, Al Karaki A, Al-Kadash A, Lobna Gharaibeh M, Ma'mon H, Aljabali AAA, Awidi A. siRNA: mechanism of action, challenges, and therapeutic approaches. Eur J Pharmacol. 2021;905:174178.
7. Alton EWFW, Armstrong DK, Ashby D, Bayfield KJ, Bilton D, Bloomfield EV, Christopher Boyd A, Brand J, Buchan R, Calcedo R. Repeated nebulisation of non-viral CFTR gene therapy in patients with cystic fibrosis: a randomised, double-blind, placebo-controlled, phase 2b trial. Lancet Respir Med. 2015;3:684–91.
8. An X, Martinez-Paniagua M, Rezvan A, Sefat SR, Fathi M, Singh S, Biswas S, Pourpak M, Yee C, Liu X, Varadarajan N. Single-dose intranasal vaccination elicits systemic and mucosal immunity against SARS-CoV-2. iScience. 2021;24:103037.
9. Anderson BR, Muramatsu H, Jha BK, Silverman RH, Weissman D, Karikó K. Nucleoside modifications in RNA limit activation of 2′-5′-oligoadenylate synthetase and increase resistance to cleavage by RNase L. Nucleic Acids Res. 2011;39:9329–38.
10. Artzy-Schnirman A, Hobi N, Schneider-Daum N, Guenat OT, Lehr CM, Sznitman J. Advanced in vitro lung-on-chip platforms for inhalation assays: from prospect to pipeline. Eur J Pharm Biopharm. 2019;144:11–7.
11. Baliga UK, Dean DA. Pulmonary gene delivery-realities and possibilities. Exp Biol Med (Maywood). 2021;246:260–74.
12. Bartel DP. MicroRNAs: genomics, biogenesis, mechanism, and function. Cell. 2004;116:281–97.
13. Bennett CF, Baker BF, Pham N, Swayze E, Geary RS. Pharmacology of antisense drugs. Annu Rev Pharmacol Toxicol. 2017;57:81–105.
14. Berger M, Lechanteur A, Evrard B, Piel G. Innovative lipoplexes formulations with enhanced siRNA efficacy for cancer treatment: where are we now? Int J Pharm. 2021;605:120851.
15. Bernstein E, Caudy AA, Hammond SM, Hannon GJ. Role for a bidentate ribonuclease in the initiation step of RNA interference. Nature. 2001;409:363–6.
16. Blanchard EL, Vanover D, Bawage SS, Tiwari PM, Rotolo L, Beyersdorf J, Peck HE, Bruno NC, Hincapie R, Michel F, Murray J, Sadhwani H, Vanderheyden B, Finn MG, Brinton MA, Lafontaine ER, Hogan RJ, Zurla C, Santangelo PJ. Treatment of influenza and SARS-CoV-2 infections via mRNA-encoded Cas13a in rodents. Nat Biotechnol. 2021;39:717–26.
17. Boateng E, Krauss-Etschmann S. miRNAs in Lung development and diseases. Int J Mol Sci. 2020;21
18. Boo SH, Kim YK. The emerging role of RNA modifications in the regulation of mRNA stability. Exp Mol Med. 2020;52:400–8.
19. Bramsen JB, Pakula MM, Hansen TB, Bus C, Langkjær N, Odadzic D, Smicius R, Wengel SL, Chattopadhyaya J, Engels JW. A screen of chemical modifications identifies position-specific modification by UNA to most potently reduce siRNA off-target effects. Nucleic Acids Res. 2010;38:5761–73.

20. Brune K, Frank J, Schwingshackl A, Finigan J, Sidhaye VK. Pulmonary epithelial barrier function: some new players and mechanisms. Am J Phys Lung Cell Mol Phys. 2015;308:L731–L45.
21. Bus T, Traeger A, Schubert US. The great escape: how cationic polyplexes overcome the endosomal barrier. J Mater Chem B. 2018;6:6904–18.
22. Bustamante-Marin XM, Ostrowski LE. Cilia and Mucociliary clearance, vol. 9. Cold Spring Harb Perspect Biol; 2017.
23. Carvalho TC, McConville JT. The function and performance of aqueous aerosol devices for inhalation therapy. J Pharm Pharmacol. 2016;68:556–78.
24. Chan JH, Lim S, Wong WS. Antisense oligonucleotides: from design to therapeutic application. Clin Exp Pharmacol Physiol. 2006;33:533–40.
25. Chang RYK, Chan HK. Lipid nanoparticles for the inhalation of mRNA. Nat Biomed Eng. 2021;5:949–50.
26. Cheng Q, Wei T, Farbiak L, Johnson LT, Dilliard SA, Siegwart DJ. Selective organ targeting (SORT) nanoparticles for tissue-specific mRNA delivery and CRISPR-Cas gene editing. Nat Nanotechnol. 2020;15:313–20.
27. Chery J. RNA therapeutics: RNAi and antisense mechanisms and clinical applications. Postdoc J. 2016;4:35–50.
28. Chow MYT, Chang RYK, Chan HK. Inhalation delivery technology for genome-editing of respiratory diseases. Adv Drug Deliv Rev. 2021;168:217–28.
29. Chow MYT, Qiu Y, Lam JKW. Inhaled RNA therapy: from promise to reality. Trends Pharmacol Sci. 2020;41:715–29.
30. ClinicalTrials.gov (a). Efficacy and safety of four escalating dose regimens of TPI ASM8 in patients with allergic asthma. Accessed 23/02/2023. https://beta.clinicaltrials.gov/study/NCT00822861
31. ClinicalTrials.gov (b). Safety, efficacy, PK, and PD characteristics of orally inhaled SB010 in male patients with mild asthma (multiple dose). Accessed 23/02/2023. https://beta.clinicaltrials.gov/study/NCT01743768
32. ClinicalTrials.gov (c). Study evaluating the effects of AIR645 on allergen-induced airway responses in subjects with mild atopic asthma. Accessed 23/02/2023. https://beta.clinicaltrials.gov/study/NCT00941577
33. ClinicalTrials.gov (d). Efficacy, safety, and tolerability of remlarsen (MRG-201) following intradermal injection in subjects with a history of keloids. Accessed 23/02/2023. https://beta.clinicaltrials.gov/study/NCT03601052
34. ClinicalTrials.gov (e). Intranasal ALN-RSV01 administered to adult volunteers experimentally inoculated with respiratory syncytial virus. Accessed 23/02/2023. https://beta.clinicaltrials.gov/study/NCT00496821
35. ClinicalTrials.gov (f). Phase 2b study of ALN-RSV01 in lung transplant patients infected with respiratory syncytial virus (RSV). Accessed 23/02/2023. https://beta.clinicaltrials.gov/study/NCT01065935
36. ClinicalTrials.gov (g). Safety, tolerability, and pharmacokinetic study of TRK-250 for patients with idiopathic pulmonary fibrosis. Accessed 23/02/2023. https://beta.clinicaltrials.gov/study/NCT03727802
37. Colombo S, Cun D, Remaut K, Bunker M, Zhang J, Martin-Bertelsen B, Yaghmur A, Braeckmans K, Nielsen HM, Foged C. Mechanistic profiling of the siRNA delivery dynamics of lipid-polymer hybrid nanoparticles. J Control Release. 2015;201:22–31.
38. Cone RA. Barrier properties of mucus. Adv Drug Deliv Rev. 2009;61:75–85.
39. Conte G, Costabile G, Baldassi D, Rondelli V, Bassi R, Colombo D, Linardos G, Fiscarelli EV, Sorrentino R, Miro A. Hybrid lipid/polymer nanoparticles to tackle the cystic fibrosis mucus barrier in siRNA delivery to the lungs: does PEGylation make the difference? ACS Appl Mater Interfaces. 2022;14:7565–78.
40. Costa B, Boueri B, Oliveira C, Silveira I, Ribeiro AJ. Lipoplexes and polyplexes as nucleic acids delivery nanosystems: the current state and future considerations. Expert Opin Drug Deliv. 2022;19:577–94.

41. Cryan S-A, Sivadas N, Garcia-Contreras L. In vivo animal models for drug delivery across the lung mucosal barrier. Adv Drug Deliv Rev. 2007;59:1133–51.
42. d'Angelo I, Costabile G, Durantie E, Brocca P, Rondelli V, Russo A, Russo G, Miro A, Quaglia F, Petri-Fink A, Rothen-Rutishauser B, Ungaro F. Hybrid lipid/polymer nanoparticles for pulmonary delivery of siRNA: development and fate upon in vitro deposition on the human epithelial airway barrier. J Aerosol Med Pulm Drug Deliv. 2018;31:170–81.
43. de Groot AM, Thanki K, Gangloff M, Falkenberg E, Zeng X, van Bijnen DCJ, van Eden W, Franzyk H, Nielsen HM, Broere F, Gay NJ, Foged C, Sijts A. Immunogenicity testing of lipidoids in vitro and in silico: modulating lipidoid-mediated TLR4 activation by nanoparticle design. Mol Ther Nucleic Acids. 2018;11:159–69.
44. Degors IMS, Wang C, Rehman ZU, Zuhorn IS. Carriers break barriers in drug delivery: endocytosis and endosomal escape of gene delivery vectors. Acc Chem Res. 2019;52:1750–60.
45. Ding L, Tang S, Wyatt TA, Knoell DL, Oupický D. Pulmonary siRNA delivery for lung disease: review of recent progress and challenges. J Control Release. 2021;330:977–91.
46. Duan L, Ouyang K, Xu X, Xu L, Wen C, Zhou X, Qin Z, Xu Z, Sun W, Liang Y. Nanoparticle delivery of CRISPR/Cas9 for genome editing. Front Genet. 2021;12:673286.
47. Faria J, Magalhães M, Veiga F, Santos AC, Figueiras A. Micelleplexes: A promising nanocarrier for the transport of genetic material and drugs. In: Patra JK, Shukla AC, Das G, editors. Advances in pharmaceutical biotechnology: recent progress and future applications. Springer: Singapore; 2020.
48. Farokhzad OC, Langer R. Impact of nanotechnology on drug delivery. ACS Nano. 2009;3:16–20.
49. Fire A, Xu S, Montgomery MK, Kostas SA, Driver SE, Mello CC. Potent and specific genetic interference by double-stranded RNA in Caenorhabditis elegans. Nature. 1998;391:806–11.
50. Fluiter K, Mook ORF, Baas F. The therapeutic potential of LNA-modified siRNAs: reduction of off-target effects by chemical modification of the siRNA sequence. siRNA miRNA Gene Silenc. 2009:1–15.
51. Garcia-Mouton C, Hidalgo A, Cruz A, Perez-Gil J. The lord of the lungs: the essential role of pulmonary surfactant upon inhalation of nanoparticles. Eur J Pharm Biopharm. 2019;144:230–43.
52. Gauvreau GM, Pageau R, Seguin R, Carballo D, Gauthier J, D'Anjou H, Campbell H, Watson R, Mistry M, Parry-Billings M, Killian K, Renzi PM. Dose-response effects of TPI ASM8 in asthmatics after allergen. Allergy. 2011;66:1242–8.
53. Gilleron J, Querbes W, Zeigerer A, Borodovsky A, Marsico G, Schubert U, Manygoats K, Seifert S, Andree C, Stoter M, Epstein-Barash H, Zhang L, Koteliansky V, Fitzgerald K, Fava E, Bickle M, Kalaidzidis Y, Akinc A, Maier M, Zerial M. Image-based analysis of lipid nanoparticle-mediated siRNA delivery, intracellular trafficking and endosomal escape. Nat Biotechnol. 2013;31:638–46.
54. Goodwin EC, Rottman FM. The 3′-flanking sequence of the bovine growth hormone gene contains novel elements required for efficient and accurate polyadenylation. J Biol Chem. 1992;267:16330–4.
55. Guan S, Munder A, Hedtfeld S, Braubach P, Glage S, Zhang L, Lienenklaus S, Schultze A, Hasenpusch G, Garrels W, Stanke F, Miskey C, Johler SM, Kumar Y, Tummler B, Rudolph C, Ivics Z, Rosenecker J. Self-assembled peptide-poloxamine nanoparticles enable in vitro and in vivo genome restoration for cystic fibrosis. Nat Nanotechnol. 2019;14:287–97.
56. Gunther M, Lipka J, Malek A, Gutsch D, Kreyling W, Aigner A. Polyethylenimines for RNAi-mediated gene targeting in vivo and siRNA delivery to the lung. Eur J Pharm Biopharm. 2011:1.
57. Gupta A, Andresen JL, Manan RS, Langer R. Nucleic acid delivery for therapeutic applications. Adv Drug Deliv Rev. 2021;178:113834.
58. Hall TMT. Multiple modes of RNA recognition by zinc finger proteins. Curr Opin Struct Biol. 2005;15:367–73.
59. Hastedt JE, Bäckman P, Clark AR, Doub W, Hickey A, Hochhaus G, Kuehl PJ, Lehr C-M, Mauser P, McConville J. Scope and relevance of a pulmonary biopharmaceutical classification system AAPS/FDA/USP workshop, March 16–17th, 2015 in Baltimore, MD. Springer; 2016.

60. Hou X, Zaks T, Langer R, Dong Y. Lipid nanoparticles for mRNA delivery. Nat Rev Mater. 2021:1–17.
61. Huckaby JT, Lai SK. PEGylation for enhancing nanoparticle diffusion in mucus. Adv Drug Deliv Rev. 2018;124:125–39.
62. Huotari J, Helenius A. Endosome maturation. EMBO J. 2011;30:3481–500.
63. Hussell T, Bell TJ. Alveolar macrophages: plasticity in a tissue-specific context. Nat Rev Immunol. 2014;14:81–93.
64. Hutvagner G, Zamore PD. A microRNA in a multiple-turnover RNAi enzyme complex. Science. 2002;297:2056–60.
65. Ibrahim BM, Tsifansky MD, Yang Y, Yeo Y. Challenges and advances in the development of inhalable drug formulations for cystic fibrosis lung disease. Expert Opin Drug Deliv. 2011;8:451–66.
66. Ihara D, Hattori N, Horimasu Y, Masuda T, Nakashima T, Senoo T, Iwamoto H, Fujitaka K, Okamoto H, Kohno N. Histological quantification of gene silencing by intratracheal administration of dry powdered small-interfering RNA/chitosan complexes in the murine Lung. Pharm Res. 2015;32:3877–85.
67. Imaoka H, Campbell H, Babirad I, Watson RM, Mistry M, Sehmi R, Gauvreau GM. TPI ASM8 reduces eosinophil progenitors in sputum after allergen challenge. Clin Exp Allergy. 2011;41:1740–6.
68. Ito T, Fukuhara M, Okuda T, Okamoto H. Naked pDNA/hyaluronic acid powder shows excellent long-term storage stability and gene expression in murine lungs. Int J Pharm. 2020;574:118880.
69. Jackson AL, Bartz SR, Schelter J, Kobayashi SV, Burchard J, Mao M, Li B, Cavet G, Linsley PS. Expression profiling reveals off-target gene regulation by RNAi. Nat Biotechnol. 2003;21:635–7.
70. Jain KK. Role of nanobiotechnology in drug delivery. Drug Deliv Syst. 2020:55–73.
71. Jansen EM, Frijlink HW, Hinrichs WL, Ruigrok MJ. Are inhaled mRNA vaccines safe and effective? A review of preclinical studies. Expert Opin Drug Deliv. 2022;19:1471–85.
72. Jia Y, Krishnan L, Omri A. Nasal and pulmonary vaccine delivery using particulate carriers. Expert Opin Drug Deliv. 2015;12:993–1008.
73. Jiang Y, Fan M, Yang Z, Liu X, Xu Z, Liu S, Feng G, Tang S, Li Z, Zhang Y, Chen S, Yang C, Law WC, Dong B, Xu G, Yong KT. Recent advances in nanotechnology approaches for non-viral gene therapy. Biomater Sci. 2022:1.
74. Johnson MD, Widdicombe JH, Allen L, Barbry P, Dobbs LG. Alveolar epithelial type I cells contain transport proteins and transport sodium, supporting an active role for type I cells in regulation of lung liquid homeostasis. Proc Natl Acad Sci. 2002;99:1966–71.
75. Jovic M, Sharma M, Rahajeng J, Caplan S. The early endosome: a busy sorting station for proteins at the crossroads. Histol Histopathol. 2010;25:99–112.
76. Kaczmarek JC, Kowalski PS, Anderson DG. Advances in the delivery of RNA therapeutics: from concept to clinical reality. Genome Med. 2017;9:1–16.
77. Karikó K, Muramatsu H, Welsh FA, Ludwig J, Kato H, Akira S, Weissman D. Incorporation of pseudouridine into mRNA yields superior nonimmunogenic vector with increased translational capacity and biological stability. Mol Ther. 2008;16:1833–40.
78. Ke X, Shelton L, Hu Y, Zhu Y, Chow E, Tang H, Santos JL, Mao HQ. Surface-functionalized PEGylated nanoparticles deliver messenger RNA to pulmonary immune cells. ACS Appl Mater Interfaces. 2020;12:35835–44.
79. Keil TWM, Baldassi D, Merkel OM. T-cell targeted pulmonary siRNA delivery for the treatment of asthma. Wiley Interdiscip Rev Nanomed Nanobiotechnol. 2020;12:e1634.
80. Keil TWM, Feldmann DP, Costabile G, Zhong Q, da Rocha S, Merkel OM. Characterization of spray dried powders with nucleic acid-containing PEI nanoparticles. Eur J Pharm Biopharm. 2019;143:61–9.
81. Keil TW, Zimmermann CM, Baldassi D, Adams F, Friess W, Mehta A, Merkel OM. Impact of crystalline and amorphous matrices on successful spray drying of siRNA polyplexes for inhalation of nano-in-microparticles. Adv Ther. 2021:1.

82. Kim J, Jozic A, Lin Y, Eygeris Y, Bloom E, Tan X, Acosta C, MacDonald KD, Welsher KD, Sahay G. Engineering lipid nanoparticles for enhanced intracellular delivery of mRNA through inhalation. ACS Nano. 2022;16:14792–806.
83. Knoch M, Keller M. The customised electronic nebuliser: a new category of liquid aerosol drug delivery systems. Expert Opin Drug Deliv. 2005;2:377–90.
84. Kole R, Krainer AR, Altman S. RNA therapeutics: beyond RNA interference and antisense oligonucleotides. Nat Rev Drug Discov. 2012;11:125–40.
85. Kolte A, Patil S, Lesimple P, Hanrahan JW, Misra A. PEGylated composite nanoparticles of PLGA and polyethylenimine for safe and efficient delivery of pDNA to lungs. Int J Pharm. 2017;524:382–96.
86. Krohn-Grimberghe M, Mitchell MJ, Schloss MJ, Khan OF, Courties G, Guimaraes PPG, Rohde D, Cremer S, Kowalski PS, Sun Y, Tan M, Webster J, Wang K, Iwamoto Y, Schmidt SP, Wojtkiewicz GR, Nayar R, Frodermann V, Hulsmans M, Chung A, Hoyer FF, Swirski FK, Langer R, Anderson DG, Nahrendorf M. Nanoparticle-encapsulated siRNAs for gene silencing in the haematopoietic stem-cell niche. Nat Biomed Eng. 2020;4:1076–89.
87. Kumar US, Afjei R, Ferrara K, Massoud TF, Paulmurugan R. Gold-nanostar-chitosan-mediated delivery of SARS-CoV-2 DNA vaccine for respiratory mucosal immunization: development and proof-of-principle. ACS Nano. 2021:1.
88. Lam JKW, Chow MYT, Zhang Y, Leung SWS. siRNA versus miRNA as therapeutics for gene silencing. Mol Ther Nucleic Acids. 2015;4:e252.
89. Lavorini F, Janson C, Braido F, Stratelis G, Løkke A. What to consider before prescribing inhaled medications: a pragmatic approach for evaluating the current inhaler landscape. Ther Adv Respir Dis. 2019;13:1753466619884532.
90. Leng D, Thanki K, Fattal E, Foged C, Yang M. Engineering of budesonide-loaded lipid-polymer hybrid nanoparticles using a quality-by-design approach. Int J Pharm. 2018;548:740–6.
91. Leonova EI, Gainetdinov RR. CRISPR/Cas9 technology in translational biomedicine. Cell Physiol Biochem. 2020;54:354–70.
92. Levine-Tiefenbrun M, Yelin I, Alapi H, Herzel E, Kuint J, Chodick G, Gazit S, Patalon T, Kishony R. Waning of SARS-CoV-2 booster viral-load reduction effectiveness. Nat Commun. 2022;13:1237.
93. Lin S, Staahl BT, Alla RK, Doudna JA. Enhanced homology-directed human genome engineering by controlled timing of CRISPR/Cas9 delivery. elife. 2014;3:e04766.
94. Liu Q, Guan J, Song R, Zhang X, Mao S. Physicochemical properties of nanoparticles affecting their fate and the physiological function of pulmonary surfactants. Acta Biomater. 2022;140:76–87.
95. Lokugamage MP, Vanover D, Beyersdorf J, Hatit MZC, Rotolo L, Echeverri ES, Peck HE, Ni H, Yoon J-K, Kim YT. Optimization of lipid nanoparticles for the delivery of nebulized therapeutic mRNA to the lungs. Nat Biomed Eng. 2021;5:1059–68.
96. Marczynski M, Jiang K, Blakeley M, Srivastava V, Vilaplana F, Crouzier T, Lieleg O. Structural alterations of mucins are associated with losses in functionality. Biomacromolecules. 2021;22:1600–13.
97. Martin AR, Finlay WH. Nebulizers for drug delivery to the lungs. Expert Opin Drug Deliv. 2015;12:889–900.
98. McLachlan G, Alton EWFW, Boyd AC, Clarke NK, Davies JC, Gill DR, Griesenbach U, Hickmott JW, Hyde SC, Miah KM, Molina CJ. Progress in respiratory gene therapy. Hum Gene Ther. 2022;33:893–912.
99. Merkel OM, Kissel T. Nonviral pulmonary delivery of siRNA. Acc Chem Res. 2012;45:961–70.
100. Merkel OM, Beyerle A, Librizzi D, Pfestroff A, Behr TM, Sproat B, Barth PJ, Kissel T. Nonviral siRNA delivery to the lung: investigation of PEG-PEI Polyplexes and their in vivo performance. Mol Pharm. 2009;6:1246–60.
101. Mohamed A, Pekoz AY, Ross K, Hutcheon GA, Saleem IY. Pulmonary delivery of nanocomposite microparticles (NCMPs) incorporating miR-146a for treatment of COPD. Int J Pharm. 2019;569:118524.

102. Mohwald M, Pinnapireddy SR, Wonnenberg B, Pourasghar M, Jurisic M, Jung A, Fink-Straube C, Tschernig T, Bakowsky U, Schneider M. Aspherical, nanostructured microparticles for targeted gene delivery to alveolar macrophages. Adv Healthc Mater. 2017:6.
103. Mukherjee A, Waters AK, Kalyan P, Achrol AS, Kesari S, Yenugonda VM. Lipid-polymer hybrid nanoparticles as a next-generation drug delivery platform: state of the art, emerging technologies, and perspectives. Int J Nanomedicine. 2019;14:1937–52.
104. Munir M, Kett VL, Dunne NJ, McCarthy HO. Development of a spray-dried formulation of peptide-DNA nanoparticles into a dry powder for pulmonary delivery using factorial design. Pharm Res. 2022;39:1215–32.
105. Murgia X, Loretz B, Hartwig O, Hittinger M, Lehr C-M. The role of mucus on drug transport and its potential to affect therapeutic outcomes. Adv Drug Deliv Rev. 2018;124:82–97.
106. Nemati E, Mokhtarzadeh A, Panahi-Azar V, Mohammadi A, Hamishehkar H, Mesgari-Abbasi M, Dolatabadi JEN, de la Guardia M. Ethambutol-loaded solid lipid nanoparticles as dry powder inhalable formulation for tuberculosis therapy. AAPS PharmSciTech. 2019;20:1–9.
107. Newman SP. Drug delivery to the lungs: challenges and opportunities. Ther Deliv. 2017;8:647–61.
108. Ng B, Cash-Mason T, Wang Y, Seitzer J, Burchard J, Brown D, Dudkin V, Davide J, Jadhav V, Sepp-Lorenzino L. Intratracheal administration of siRNA triggers mRNA silencing in the lung to modulate T cell immune response and lung Inflammation. Mol Ther Nucleic Acids. 2019;16:194–205.
109. Nguyen J, Steele TWJ, Merkel O, Reul R, Kissel T. Fast degrading polyesters as siRNA nano-carriers for pulmonary gene therapy. J Control Release. 2008;132:243–51.
110. Nimesh S, Chandra R, Gupta N. Advances in nanomedicine for the delivery of therapeutic nucleic acids, Woodhead Publishing; 2017.
111. Okamoto H, Shiraki K, Yasuda R, Danjo K, Watanabe Y. Chitosan-interferon-beta gene complex powder for inhalation treatment of lung metastasis in mice. J Control Release. 2011;150:187–95.
112. Okuda T, Morishita M, Mizutani K, Shibayama A, Okazaki M, Okamoto H. Development of spray-freeze-dried siRNA/PEI powder for inhalation with high aerosol performance and strong pulmonary gene silencing activity. J Control Release. 2018;279:99–113.
113. Pardi N, Tuyishime S, Muramatsu H, Kariko K, Mui BL, Tam YK, Madden TD, Hope MJ, Weissman D. Expression kinetics of nucleoside-modified mRNA delivered in lipid nanoparticles to mice by various routes. J Control Release. 2015;217:345–51.
114. Parhiz H, Shuvaev VV, Pardi N, Khoshnejad M, Kiseleva RY, Brenner JS, Uhler T, Tuyishime S, Mui BL, Tam YK, Madden TD, Hope MJ, Weissman D, Muzykantov VR. PECAM-1 directed re-targeting of exogenous mRNA providing two orders of magnitude enhancement of vascular delivery and expression in lungs independent of apolipoprotein E-mediated uptake. J Control Release. 2018;291:106–15.
115. Patel AK, Kaczmarek JC, Bose S, Kauffman KJ, Mir F, Heartlein MW, DeRosa F, Langer R, Anderson DG. Inhaled nanoformulated mRNA polyplexes for protein production in lung epithelium. Adv Mater. 2019;31:e1805116.
116. Patel V, Bardoliwala D, Lalani R, Patil S, Ghosh S, Javia A, Misra A. Development of a dry powder for inhalation of nanoparticles codelivering cisplatin and ABCC3 siRNA in lung cancer. Ther Deliv. 2021;12:651–70.
117. Paul D, Miller MH, Born J, Samaddar S, Ni H, Avila H, Krishnamurthy VR, Thirunavukkarasu K. The promising therapeutic potential of oligonucleotides for pulmonary fibrotic diseases. Expert Opin Drug Discov. 2023;18:193–206.
118. Popowski KDB, de Juan Abad L, George A, Silkstone D, Belcher E, Chung J, Ghodsi A, Lutz H, Davenport J, Flanagan M, Piedrahita J, Dinh PC, Cheng K. Inhalable exosomes outperform liposomes as mRNA and protein drug carriers to the lung. Extracell Vesicle. 2022;1:100002.
119. Popowski KD, Moatti A, Scull G, Silkstone D, Lutz H, de Juan Abad BL, George A, Belcher E, Zhu D, Mei X, Cheng X, Cislo M, Ghodsi A, Cai Y, Huang K, Li J, Brown AC, Greenbaum A, Dinh PC, Cheng K. Inhalable dry powder mRNA vaccines based on extracellular vesicles. Matter. 2022;5:2960–74.

120. Qiu Y, Man RCH, Liao Q, Kung KLK, Chow MYT, Lam JKW. Effective mRNA pulmonary delivery by dry powder formulation of PEGylated synthetic KL4 peptide. J Control Release. 2019;314:102–15.
121. Raesch SS, Tenzer S, Storck W, Rurainski A, Selzer D, Ruge CA, Perez-Gil J, Schaefer UF, Lehr C-M. Proteomic and lipidomic analysis of nanoparticle corona upon contact with lung surfactant reveals differences in protein, but not lipid composition. ACS Nano. 2015;9:11872–85.
122. Roberts TC, Langer R, Wood MJA. Advances in oligonucleotide drug delivery. Nat Rev Drug Discov. 2020;19:673–94.
123. Ross NL, Munsell EV, Sabanayagam C, Sullivan MO. Histone-targeted polyplexes avoid endosomal escape and enter the nucleus during postmitotic redistribution of ER membranes. Mol Ther Nucleic Acids. 2015;4:e226.
124. Rothen-Rutishauser B, Blank F, Mühlfeld C, Gehr P. In vitro models of the human epithelial airway barrier to study the toxic potential of particulate matter. Expert Opin Drug Metab Toxicol. 2008;4:1075–89.
125. Rotolo L, Vanover D, Bruno NC, Peck HE, Zurla C, Murray J, Noel RK, O'Farrell L, Arainga M, Orr-Burks N, Joo JY, Chaves LCS, Jung Y, Beyersdorf J, Gumber S, Guerrero-Ferreira R, Cornejo S, Thoresen M, Olivier AK, Kuo KM, Gumbart JC, Woolums AR, Villinger F, Lafontaine ER, Hogan RJ, Finn MG, Santangelo PJ. Species-agnostic polymeric formulations for inhalable messenger RNA delivery to the lung. Nat Mater. 2022:1.
126. Ruge CA, Kirch J, Lehr C-M. Pulmonary drug delivery: from generating aerosols to overcoming biological barriers – therapeutic possibilities and technological challenges. Lancet Respir Med. 2013;1:402–13.
127. Rupaimoole R, Slack FJ. MicroRNA therapeutics: towards a new era for the management of cancer and other diseases. Nat Rev Drug Discov. 2017;16:203–22.
128. Sahin U, Karikó K, Türeci Ö. mRNA-based therapeutics—developing a new class of drugs. Nat Rev Drug Discov. 2014;13:759–80.
129. Samaridou E, Heyes J, Lutwyche P. Lipid nanoparticles for nucleic acid delivery: current perspectives. Adv Drug Deliv Rev. 2020;154-155:37–63.
130. Schaffert D, Wagner E. Gene therapy progress and prospects: synthetic polymer-based systems. Gene Ther. 2008;15:1131–8.
131. Schlich M, Palomba R, Costabile G, Mizrahy S, Pannuzzo M, Peer D, Decuzzi P. Cytosolic delivery of nucleic acids: the case of ionizable lipid nanoparticles. Bioeng Transl Med. 2021;6:e10213.
132. Scott CC, Vacca F, Gruenberg J. Endosome maturation, transport and functions. Semin Cell Dev Biol. 2014;31:2–10.
133. Sigurdsson HH, Kirch J, Lehr CM. Mucus as a barrier to lipophilic drugs. Int J Pharm. 2013;453:56–64.
134. Sioud M, Furset G, Cekaite L. Suppression of immunostimulatory siRNA-driven innate immune activation by 2′-modified RNAs. Biochem Biophys Res Commun. 2007;361:122–6.
135. Stewart MP, Sharei A, Ding X, Sahay G, Langer R, Jensen KF. In vitro and ex vivo strategies for intracellular delivery. Nature. 2016;538:183–92.
136. Tagalakis AD, Munye MM, Ivanova R, Chen H, Smith CM, Aldossary AM, Rosa LZ, Moulding D, Barnes JL, Kafetzis KN. Effective silencing of ENaC by siRNA delivered with epithelial-targeted nanocomplexes in human cystic fibrosis cells and in mouse lung. Thorax. 2018;73:847–56.
137. Tam A, Kulkarni J, An K, Li L, Dorscheid DR, Singhera GK, Bernatchez P, Reid G, Chan K, Witzigmann D, Cullis PR, Sin DD, Lim CJ. Lipid nanoparticle formulations for optimal RNA-based topical delivery to murine airways. Eur J Pharm Sci. 2022;176:106234.
138. Torras N, García-Díaz M, Fernández-Majada V, Martínez E. Mimicking epithelial tissues in three-dimensional cell culture models. Front Bioeng Biotechnol. 2018;6:197.
139. Trubetskoy VS, Griffin JB, Nicholas AL, Nord EM, Zhao X, Peterson RM, Wooddell CI, Rozema DB, Wakefield DH, Lewis DL. Phosphorylation-specific status of RNAi triggers in pharmacokinetic and biodistribution analyses. Nucleic Acids Res. 2017;45:1469–78.

140. Tsukamoto M, Okuda T, Okamoto H, Higuchi Y, Kawakami S, Yamashita F, Hashida M. Bovine serum albumin as a lyoprotectant for preparation of DNA dry powder formulations using the spray-freeze drying method. Biol Pharm Bull. 2012;35:1178–81.
141. ur Rehman Z, Hoekstra D, Zuhorn IS. Mechanism of polyplex- and lipoplex-mediated delivery of nucleic acids: real-time visualization of transient membrane destabilization without endosomal lysis. ACS Nano. 2013;7:3767–77.
142. Vencken S, Foged C, Ramsey JM, Sweeney L, Cryan S-A, MacLoughlin RJ, Greene CM. Nebulised lipid–polymer hybrid nanoparticles for the delivery of a therapeutic anti-inflammatory microRNA to bronchial epithelial cells. ERJ Open Res. 2019;5
143. Vermeulen LMP, Brans T, Samal SK, Dubruel P, Demeester J, De Smedt SC, Remaut K, Braeckmans K. Endosomal size and membrane leakiness influence proton sponge-based rupture of endosomal vesicles. ACS Nano. 2018;12:2332–45.
144. Wang C, Zhang Y, Dong Y. Lipid nanoparticle-mRNA formulations for therapeutic applications. Acc Chem Res. 2021;54:4283–93.
145. Wang Z, Popowski KD, Zhu D, de Juan Abad BL, Wang X, Liu M, Lutz H, De Naeyer N, DeMarco CT, Denny TN, Dinh P-UC, Li Z, Cheng K. Exosomes decorated with a recombinant SARS-CoV-2 receptor-binding domain as an inhalable COVID-19 vaccine. Nat Biomed Eng. 2022;6:791–805.
146. Watts JK, Deleavey GF, Damha MJ. Chemically modified siRNA: tools and applications. Drug Discov Today. 2008;13:842–55.
147. Weber S, Zimmer A, Pardeike J. Solid lipid nanoparticles (SLN) and nanostructured lipid carriers (NLC) for pulmonary application: a review of the state of the art. Eur J Pharm Biopharm. 2014;86:7–22.
148. Winkeljann B, Keul DC, Merkel OM. Engineering poly- and micelleplexes for nucleic acid delivery – a reflection on their endosomal escape. J Control Release. 2023;353:518–34.
149. Wisnewski AV, Campillo Luna J, Redlich CA. Human IgG and IgA responses to COVID-19 mRNA vaccines. PLoS One. 2021;16:e0249499.
150. Wittekindt OH. Tight junctions in pulmonary epithelia during lung inflammation. Pflügers Archiv-Eur J Physiol. 2017;469:135–47.
151. Wu J, Wu L, Wan F, Rantanen J, Cun D, Yang M. Effect of thermal and shear stresses in the spray drying process on the stability of siRNA dry powders. Int J Pharm. 2019;566:32–9.
152. Xu Y, Harinck L, Lokras AG, Gerde P, Selg E, Sjoberg CO, Franzyk H, Thakur A, Foged C. Leucine improves the aerosol performance of dry powder inhaler formulations of siRNA-loaded nanoparticles. Int J Pharm. 2022;621:121758.
153. Xu Y, Thakur A, Zhang Y, Foged C. Inhaled RNA therapeutics for obstructive airway diseases: recent advances and future prospects. Pharmaceutics. 2021;13
154. Yang J, Arya S, Lung P, Lin Q, Huang J, Li Q. Hybrid nanovaccine for the co-delivery of the mRNA antigen and adjuvant. Nanoscale. 2019;11:21782–9.
155. Yang W, Yan J, Zhuang P, Ding T, Chen Y, Zhang Y, Zhang H, Cui W. Progress of delivery methods for CRISPR-Cas9. Expert Opin Drug Deliv. 2022;19:913–26.
156. Zhang H, Leal J, Soto MR, Smyth HDC, Ghosh D. Aerosolizable lipid nanoparticles for pulmonary delivery of mRNA through design of experiments. Pharmaceutics. 2020;12
157. Zhang Y, Sun C, Wang C, Jankovic KE, Dong Y. Lipids and lipid derivatives for RNA delivery. Chem Rev. 2021;121:12181–277.
158. Zhao Z, Anselmo AC, Mitragotri S. Viral vector-based gene therapies in the clinic. Bioeng Transl Med. 2022;7:e10258.
159. Zimmermann CM, Baldassi D, Chan K, Adams NBP, Neumann A, Porras-Gonzalez DL, Wei X, Kneidinger N, Stoleriu MG, Burgstaller G. Spray drying siRNA-lipid nanoparticles for dry powder pulmonary delivery. J Control Release. 2022;351:137–50.

Intranasal and Inhaled Vaccines

Michael Yee-Tak Chow and Jenny Ka Wing Lam

Abstract Many respiratory infections have epidemic and pandemic potential, as evidenced by influenza and the recent COVID-19 pandemic. Vaccination is an established and effective strategy to contain the spread of communicable diseases. Most vaccines are injectables that induce potent systemic immunity. However, they are less efficient in eliciting immune responses at the respiratory mucosa, the site where many respiratory pathogens invade and replicate. This therapeutic inadequacy may be addressed by intranasal and inhaled vaccines which can trigger mucosal immunity. Only limited intranasal and inhaled vaccines are clinically available, including a live-attenuated influenza vaccine and two recently marketed viral vector vaccines targeting SARS-CoV-2. Nevertheless, the immunological advantages of mucosal vaccines should not be undermined, which include the activation of tissue-specific mucosal immune response and common mucosal immune system, while they also induce systemic immunity as injectable vaccines do. In this chapter, the mechanisms of intranasal and inhaled vaccines are introduced and their advantages over injectable vaccines are discussed. Other social and economic benefits of intranasal and inhaled vaccines are also presented, followed by an overview of associated challenges and limitations that must be overcome and addressed before their therapeutic potential can be better utilised.

Keywords Aerosolised vaccine · Dry powder inhaler · Epidemic · Influenza · Mucosal immunity · Nebuliser · Respiratory infections

M. Y.-T. Chow
Department of Pharmaceutics, UCL School of Pharmacy, University College London, London, UK

Department of Pharmacology and Pharmacy, LKS Faculty of Medicine, The University of Hong Kong, Pokfulam, Hong Kong

J. K. W. Lam (✉)
Department of Pharmaceutics, UCL School of Pharmacy, University College London, Brunswick Square, London, UK
e-mail: jenny.lam@ucl.ac.uk

J. K. W. Lam, P. C. L. Kwok (eds.), *Respiratory Delivery of Biologics, Nucleic Acids, and Vaccines*, AAPS Introductions in the Pharmaceutical Sciences 8,
https://doi.org/10.1007/978-3-031-47567-2_6

1 Background

Emerging pathogens continually pose serious health challenges at both individual and societal levels in the form of communicable diseases, which at times escalate to epidemics or even pandemics. Many of the recent epidemics and pandemics affect the lungs, including tuberculosis [1], influenza [2], and COVID-19 [3, 4]. In the face of these threats to global health, vaccination has been and remains a powerful strategy to contain or mitigate the spread of communicable diseases.

Vaccination is an old concept that dates back to the fifteenth century or even earlier [5]. It was popularised by the British physician Edward Jenner in the late 1700s, who successfully prevented smallpox infection by inoculation with cowpox viruses [6]. Fundamentally, vaccination exposes the recipient to an antigen, which is derived from the pathogen, in order to trigger an immune response [7]. If the antigen is encountered for the first time, the innate immune response is activated. Innate immunity is non-specific and mediated through cellular response involving leukocytes (e.g., mast cells, natural killer cells, and eosinophils) as well as phagocytes (e.g., macrophages, neutrophils, and dendritic cells). Some of these cells, such as macrophages and dendritic cells, can process phagocytised antigens, display them on their surfaces and present them to lymphocytes such as T cells and B cells. This marks the activation of the adaptive (or acquired) immune response and confers immunological memory. Adaptive immunity is highly specific and is capable of rapidly initiating profound cellular and humoral immunity when the same antigen is reencountered. The former involves $CD8^+$ cytotoxic T cells that cause apoptosis in infected cells, while the latter involves B cells and their differentiated effector cells, plasma cells, that secret antigen-specific antibodies, with both responses being modulated by $CD4^+$ T helper cells. Vaccination allows the immune system to be primed with new antigens in a regulated and safe manner, thereby conferring protection against the pathogen (Fig. 1).

Since the discovery of vaccines two centuries ago, different vaccine platforms have been developed (Table 1) [8]. Traditional vaccine platforms are live-attenuated vaccines and inactivated vaccines. Live-attenuated vaccines are viable virus particles that are modified to reduce their pathogenicity. A notable example is FluMist® (and subsequently FluMist® Quadrivalent; sold as Fluenz® Tetra in the European Union), which is an intranasal live-attenuated influenza vaccine developed by MedImmune, LLC (then acquired by AstraZeneca). Apart from being attenuated, the virus particles in FluMist® are also engineered to be cold-adapted and temperature-sensitive. They can efficiently replicate at a colder temperature of 25 °C at which the replication of other wild-type influenza strains is impaired, but they have restricted replication at body temperature [9]. As live-attenuated vaccines retain the capacity to exhibit the full viral replication cycle, they are not suitable for individuals with compromised immunity. This is distinct from inactivated vaccines, which are 'killed' or destroyed using chemicals, heat, or radiation and their infectivity is lost as a result. Both platforms allow the immune system to detect the whole virus particle and help induce strong and broad immune responses [10, 11].

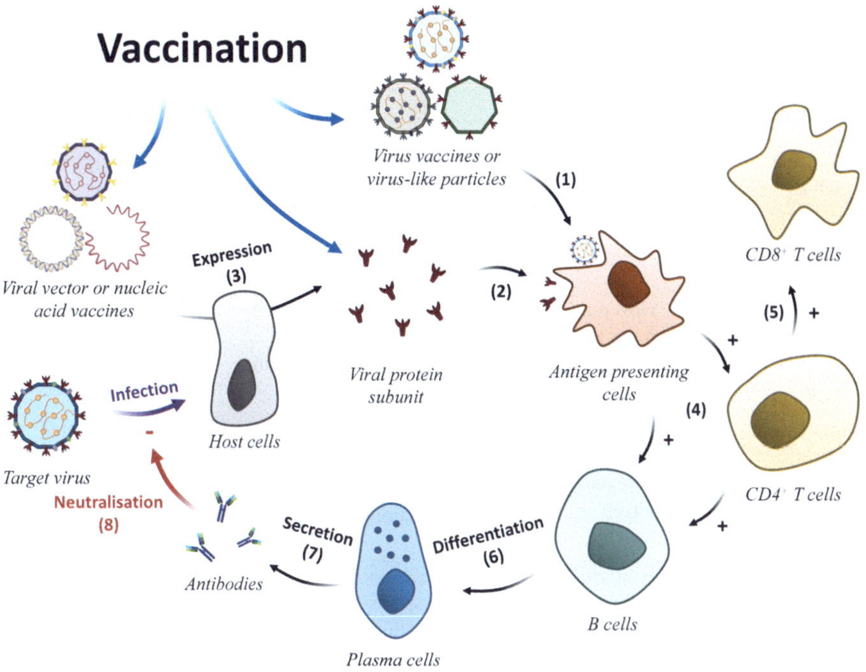

Fig. 1 Schematic diagram of the action of vaccines. Traditional vaccine platforms (live-attenuated vaccines and inactivated vaccines) and virus-like particle vaccines express antigens on their surfaces which can be detected by antigen-presenting cells (APC), such as dendritic cells and pulmonary macrophages (1). Antigenic viral proteins can also be directly administered as protein subunit vaccines (2). Alternatively, vaccine platforms such as viral vector vaccines or nucleic acid (plasmid DNA and mRNA) vaccines rely on host cells for the synthesis and expression of antigens through transduction (3). APCs process and present processed antigens to lymphocytes like $CD4^+$ T helper cells and B cells (4). The former activates other lymphocytes including $CD8^+$ cytotoxic T cells and mediates immune response against the antigen (5), while the latter differentiate into plasma cells (6) that secret antibodies (7) for virus neutralisation (8) and other protection mechanisms [171]. Upon intranasal or inhaled vaccination, some of the activated T cells develop into tissue-resident memory T (T_{RM}) cells to elicit local immune responses

Virus-like particles, in contrast, only consist of the essential structural viral proteins for the formation of a particle on which the antigenic proteins attach. Other viral genomes and non-structural proteins are absent, making them different from a true virus. Subunit vaccines contain the antigen of interest derived from the targeted pathogen. They are considered to be a safer platform because they do not carry any viral genome and lack replication capacity. Examples of approved subunit vaccines include hepatitis B vaccines, which contain the hepatitis B surface antigen (HBsAg) [12], and the COVID-19 vaccine Nuvaxovid™, which comprises of SARS-CoV-2 spike (S) protein [13].

All the above platforms deliver antigens directly to the recipient. Another vaccination strategy is to deliver a viral genome that codes for the antigens to the recipient, harnessing the host cellular machinery in synthesising the antigens [14]. Viral

Table 1 Comparison of different vaccine platforms (using a virus as an example for illustration) [8, 172, 173]

Vaccine platform	Descriptions	Pros	Cons
Live-attenuated 	Live virus particles (capable of exhibiting full virus replication cycle) with modifications to reduce pathogenicity	Strong immunogenicity as the full virus is exposed to the immune system	Not suitable in individuals with compromised immunity
		Longer-lasting immune response	Risk of microbe reactivation
Inactivated 	'Killed' virus particles that have been inactivated using chemicals, heat, or radiation	Strong immunogenicity as the full virus is exposed to the immune system	Adjuvants may be required to enhance the immune response
Virus-like particle 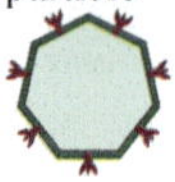	Consists of essential structural proteins for the formation of a virus particle on which virus antigens attach	Efficient cellular entry Carries no genetic materials, thus no risk of infection	Adjuvants are often needed to enhance immunogenicity
Subunit 	Comprises of viral antigens which are typically proteins [12, 13]	Carries no genetic materials, thus no risk of infection	Adjuvants are often needed to enhance immunogenicity
Viral vector 	Contains low pathogenic viruses as vectors whose viral genomes have been modified to encode the antigenic proteins [15]	Can be either replicating or non-replicating Efficient cellular entry	Prior exposure to the viral vector may compromise immunogenicity
DNA 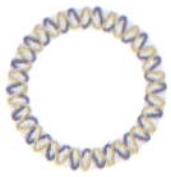	Plasmid (circular) DNA containing a genomic sequence that transcribes and translates the antigenic proteins	Sequence can be adapted to express different antigens	A delivery vector must be used
		No risk of infection	Cellular and nuclear entry can be challenging
mRNA 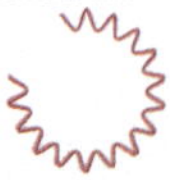	Messenger RNA containing genomic sequence that translates to the antigenic proteins [17]	Sequence can be adapted to express different antigens	A delivery vector must be used
		No risk of infection	Cellular entry can be challenging

vector vaccines adapt virus of low virulent (e.g., adenovirus used in the COVID-19 vaccine Vaxzevria™/Covishield™ co-developed by Oxford University and AstraZeneca) as a vector to deliver genes that encode the target antigens [15]. Viral vector vaccines can either retain their replication ability or have the viral replication genes deleted for improved safety. The antigen-encoding genes can also be delivered as plasmid DNA or messenger RNA (mRNA) using non-viral vectors. Notably, two of the injectable COVID-19 vaccines, tozinameran (Comirnaty® from Pfizer and BioNTech) and elasomeran (Spikevax™ from Moderna), belong to mRNA vaccines. Both of them use lipid nanoparticles (LNP) as the delivery vector [16, 17].

The existence of many vaccine platforms implies that vaccines do not represent a single class of biomolecule but can take the form of full microorganisms, proteins, or nucleic acids. Each of the platforms thus demands its own specific requirements for intranasal and inhaled formulations. Discussion on the delivery and formulation of respective biologics is covered in other chapters of this book. Currently, most vaccines are parenteral injections intended to induce strong systemic immunity. There are only a handful of approved vaccines that are administered through the mucosal route (Table 2) [18, 19]. While approved intranasal and inhaled vaccines might be lacking in number, they possess important immunological advantages over injectable vaccines, as illustrated in this chapter. Other social and economic benefits of these vaccines are also presented, followed by an overview of their challenges and limitations.

2 Principles of Intranasal and Inhaled Vaccines

Intranasal and inhaled vaccines are vaccines that are administered through the intranasal route into the nose and pulmonary route into the lung, respectively (Fig. 2). In addition to being non-invasive, the rationale of intranasal and inhaled vaccines in protecting us from respiratory infectious diseases arises from their capability to trigger immune responses that closely resemble natural infection [18, 20]. Particularly, mucosal immunity are induced. It refers to the immune responses offered by the

Table 2 Examples of clinically available mucosal vaccines

Vaccine	Targeted pathogen	Vaccine platform	Route of administration (dosage form)
Biopolio®, mOPV, and tOPV	Poliovirus	Live attenuated	Oral (liquid)
RotaTeq®	Rotavirus	Live reassortant	Oral (liquid)
Rotarix™	Rotavirus	Live attenuated	Oral (liquid)
Typhi Vivotif®	*Salmonella typhimurium*	Live attenuated	Oral (capsule)
Dukoral®	*Vibrio cholerae*	Inactivated	Oral (liquid)
Euvichol®/ShanChol™	*Vibrio cholerae*	Inactivated	Oral (liquid)
Vaxchora®	*Vibrio cholerae*	Live attenuated	Oral (liquid)
FluMist®	Influenza A & B	Live attenuated	Intranasal (liquid spray)
iNCOVACC®	SARS-CoV-2	Viral vector	Intranasal (liquid spray)
Convidecia® air	SARS-CoV-2	Viral vector	Intratracheal (liquid for nebulisation) [87]

Adapted from Refs. [18, 19]
mOPV monovalent oral polio vaccine, *tOPV* trivalent oral polio vaccine

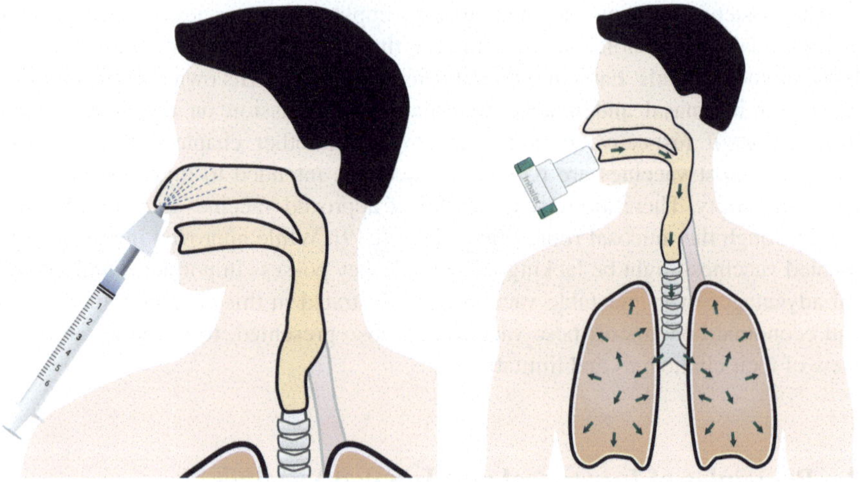

Fig. 2 Intranasal administration (left) using an intranasal atomization device and intratracheal administration (right) using a unit-dose dry powder inhaler

Table 3 Summary of major responses of mucosal immunity

Responses	Features
Tissue-specific mucosal immune response [124]	Formation of iBALT
	Humoral and cellular immune responses are mediated through local production of sIgA and recruitment of T_{RM} $CD8^+$ and $CD4^+$ cells, respectively
Common mucosal immune system	Mucosal immunity can be induced at sites other than that initially exposed to the antigens [55]
Systemic immunogenicity [65]	Serum-neutralising IgG has been detected upon mucosal vaccination, resembling the course of a natural infection
	Seroconversion has been observed in different respiratory infections such as influenza, RSV infection, and COVID-19

iBALT inducible bronchus-associated lymphoid tissue, *Ig* immunoglobulin, *RSV* respiratory syncytial virus, *sIgA* secretory IgA, T_{RM} tissue-resident memory

immune system localised at the mucosae that comprise of mucosa-associated lymphoid tissue (MALT), lymphocytes, as well as effector molecules including cytokines, chemokines and antibodies (or immunoglobulins, Ig) that function independently of the systemic immunity [20]. Mucosal immunity is inadequately induced by injectable vaccines [21, 22]. Activation of the mucosal immune system results in three critical immune responses, namely tissue-specific mucosal immune responses, common mucosal immune system, and systemic immunogenicity (Table 3) [23]. These enhanced effector and immune responses induced through direct exposure of antigens to the mucosae allow intranasal and inhaled vaccines to provide better protection than injectable vaccines [24], as elaborated below.

2.1 *Tissue-Specific Mucosal Immune Response*

The bronchus-associated lymphoid tissue (BALT) is usually absent in the lungs of healthy adults [25]. It can be induced by infection [26, 27] or inflammation due to exogenous stimuli such as cigarette smoke [28] and diesel exhaust [29]. The formation of inducible BALT (iBALT) forms the basis of mucosal immune response. iBALT is a tertiary lymphoid structure that neither demonstrates a well-defined structure, nor situates at specific anatomical locations. Rather, iBALT can be found throughout the lung and it has a diverse composition, ranging from isolated B cells clusters that lack obvious organisation [30] to fully developed lymphoid tissues that consist of B cells, T cells, and other antigen-presenting cells (APCs) including dendritic cells and macrophages [31]. Hence, iBALT contributes to both humoral and cellular immune responses. The development and organisation of iBALT is regulated under homeostatic chemokines primarily through the IL-17 pathway of inflammation, in which the expression of CXCL13 attracts B cells, and CCL19 and CCL21 attract T cells [32]. Upon the formation of iBALT, antigen acquisition and presentation can be achieved by Microfold (M) cells which are often located at the nearby epithelium [33], via the afferent lymphatics, or through the migration of antigen-acquiring dendritic cells from the airways [34]. Detail on the formation and physiology of iBALT has been recently reviewed [31].

B cells are the major lymphocytes involved in the humoral immune response. After being presented with processed antigens through major histocompatibility complexes (MHC) by the APCs, B cells begin to differentiate into plasma cells that are responsible for the secretion of Ig. Whilst IgG is the main class of Ig produced in systemic immune response, stimulation of mucosal immunity initiates the local production of a high level of secretory mucosal IgA (sIgA) in addition to serum IgG [35]. sIgA is pivotal in facilitating passive immune protection at mucosal surfaces since the mucosal secretion contains only a low level of complement proteins and leucocytes which are responsible for mediating strong immune effector functions [36]. sIgA offers a wide array of protections. Foremost, it elicits an immediate immune reaction by neutralisation, directly preventing pathogens from interacting with receptors expressed on the host cells through steric hindrance (immune exclusion) [37]. The protective effects of sIgA are not confined to the mucosal surface. Antigens that have crossed the epithelium and reached the lamina propria, and those that are inside the infected cells can also be neutralised by sIgA (intracellular neutralisation). These bound antigens are then expelled to the airways via the polymeric immunoglobulin receptors (pIgR) on the basolateral side of epithelial cells (antigen excretion) [38]. sIgA also mediates cellular immune response, including phagocytosis and cytokine release [39], through binding with the IgA Fc receptor FcαRI (also known as CD89) expressed in myeloid lineage cells like monocytes, macrophages, and neutrophils [40–42]. Efficient induction and secretion of IgA into airway lumens through mucosal vaccination is a characteristic distinction to injectable vaccines [43, 44], with the former being more effective in alleviating disease progression during the early phase of infection. Compared to sIgA, systemic antibodies are less effective in preventing systemic invasion of pathogens or abolishing disease transmission that occurs in the respiratory system [45, 46].

The induction of antigen-specific tissue-resident memory T (T_{RM}) cells represents another significant immune response unique to mucosal immunisation and constitutes the cellular component of an adaptive immune response. Pulmonary T_{RM} cells are critical in the prevention of respiratory viral infection by eliciting specialised and rapid effector responses in adaptive immunity, including the expression of inflammatory cytokines and chemokines for the enhancement of tissue antiviral resistance and recruitment of auxiliary immune cells such as dendritic cells [47]. Specifically, the $CD8^+$ T_{RM} cells can react quickly upon reencountering the antigens or pathogens [48] while the helper $CD4^+$ T_{RM} cells are engaged in the formation of protective respiratory B cells and $CD8^+$ T cells in the lung [49]. These T_{RM} cells were once considered lasting memory cells that confer enduring mucosal immunity [50]. Recent studies have been suggesting that $CD8^+$ T_{RM} cells in the lung are more short-lived than previously thought, due to the lack of adhesion molecules CD69 and CD103, with a half-life as short as 14 days and their population substantially declined within 100–200 days. This contrasts with their counterparts in other tissues such as the skin and the intestine [51, 52]. Nonetheless, $CD4^+$ T resident helper cells may still contribute to long-term immunity that initiates a rapid response to subsequent infections [53]. A booster dose in sustaining an effective memory cell population in the lungs has been suggested [18], and additional research is warranted.

2.2 Common Mucosal Immune System

The common mucosal immune system is another major and exploitable characteristic of mucosal immune response [54]. Although the initial priming of mucosal immune response can be anatomically confined to a specific mucosal tissue, it has been shown that exposure to antigens at one mucosa can trigger immunity at the distinct mucosal regions, sometimes even in different physiological systems, for example between respiratory and digestive systems [55]. This property of cross-mucosal immunity opens the possibility of achieving target mucosal immunity that is distant from the site of vaccination. In the context of intranasal and inhaled vaccines, their immunogenicity could thus rely less on the site of vaccine deposition and antigen uptake, due to common mucosal co-stimulation. In a deposition study using a cotton rat model, the whole inactivated influenza virus vaccine was administered intratracheally either to the deep lung region as liquid or to the trachea and central lung region as powders [56]. While serum IgG levels were similar between the two groups and were comparable to the positive control administered intramuscularly, indicating priming of systemic immune response, vaccine-specific IgA was not only detected in the lung but also in nasal washes. The latter was ascribed to antigen-specific B cells migrating from the site of induction to distant mucosal lymphoid tissue. In a follow-up study, inhalable powder formulations containing both the influenza subunit vaccine and HBsAg were administered to mice intratracheally, using two devices with different deposition profiles [57]. Both the lung washes and serum displayed comparable titre levels of IgA and IgG, respectively, against influenza regardless of the administration device. Intriguingly, only animals with antigens reaching the peripheral regions expressed hepatitis B-specific IgG in serum. It was proposed that airborne pathogens exhibiting specific molecular patterns can be recognised along the respiratory tract, thus rendering the induction of immune responses less dependent on the site of antigen deposition. While there is no specific receptor in the airway targeted by hepatitis B, the deposition of antigens in the deep lung region could allow sufficient time to trigger immune responses. Besides, HBsAg could form virus-like particles that passively diffuse from alveoli to draining lymph nodes, inducing immune reactions [57]. In a separate study that evaluated the effects of different sites of pulmonary deposition through varying the route of delivery (intranasal or intratracheal administration), the volume of delivered solution and the angle of tilt, a monovalent influenza vaccine was administered to BALB/c mice [58]. While similar IgA titres were detected in the nasal washes of all groups regardless of the delivery methods, there was a 1-log reduction in IgA titre in broncho-alveoli lavage for animals inoculated by the intranasal route. No difference was observed otherwise between groups that targeted deposition in different regions within the lungs. Nevertheless, only the deep lung targeted groups could present similar serum IgG titres to the control intramuscular group. Recently, a comparative study exploring the relationship between the biodistribution of vaccine and their immunogenicity in the lung using murine models and adenovirus-vectored vaccine was reported [59]. Mice vaccinated via the intratracheal route exhibited more pronounced immune responses and protective effects in the lungs than the intranasal route. These studies clearly demonstrated some important distinctions between intranasal and inhaled vaccines that call for further investigations, especially clinical studies.

2.3 Systemic Immunogenicity

The capacity of intranasal and inhaled vaccines to trigger systemic immunity in addition to mucosal immune responses conceivably makes them more advantageous than their injectable counterparts. Indeed, preclinical studies have reported that strong systemic immunity comparable to intramuscular vaccination could be induced by intranasal or inhaled vaccines [60–63], including the aforementioned deposition studies [56–58]. For instance, intratracheal administration of an influenza vaccine containing haemagglutinin antigen significantly induced robust serum antibodies as well as local IgG and IgA in mice [43]. In another study that aimed to develop an inhalable influenza powder vaccine, both powder and reconstituted vaccines administered by pulmonary route to rats induced similar serum IgG levels and higher serum IgA levels as compared to that by subcutaneous injection [64]. The inhaled vaccines could also trigger a much higher mucosal IgA and to a lesser extent IgG without causing acute histological toxicity to the lung.

Systemic immune responses triggered by intranasal and inhaled mucosal vaccination are also observed clinically, including the intranasal vaccine FluMist® among other vaccine candidates [65]. FluMist® was approved by the U.S. Food and Drug Administration (FDA) and European Medicines Agency (EMA) in 2003 and 2012, respectively [66]. It is currently indicated for the prevention of influenza disease caused by influenza type A and B in individuals of age between 2 and 49 years old. As the vaccine effectiveness appeared to decline over time, possibly due to the combined effects of antigenic drift and inter-strain competition in multivalent preparations [11, 67, 68], novel mucosal influenza vaccine candidates with improved and sustained protection are still in demand. For instance, intranasal inoculation of live-attenuated H5N1 vaccine that lacked the highly virulent non-structural protein 1 (delNS1-H5N1) was well-tolerated and able to trigger profound vaccine-specific serum Ig titres [69]. Seroconversion was observed in 75% of the subjects even after one dose of vaccine, and sIgA responses were demonstrated in 42% of participants upon completing the scheduled second vaccination. The safety and immunogenicity of another haemagglutinin antigen-based intranasal influenza vaccine candidate adjuvanted with an *E. coli*-derived heat-labile enterotoxin was evaluated in a phase II trial [70]. While the titres of induced serum anti-HA Ig were comparable among the treatment groups (with or without adjuvants), only the adjuvanted vaccines were shown to induce substantial levels of sIgA without significant differences in adverse events profile after the second dose. In another study, a single intranasal dose of M2SR (M2-deficient single replication) H3N2 influenza vaccine was shown to offer enhanced mucosal and systemic immunity [68]. Compared to the 10% rate of H3N2 seroconversion achieved by FluMist® in seronegative subjects, the M2SR vaccine candidates led to a rise of serum antibody titre in 71% of the recipients, highlighting a possible improvement in disease control.

Besides influenza, systemic and mucosal immunogenicity upon intranasal and inhaled vaccination have also been observed for other respiratory viral infections in clinical trials, notably respiratory syncytial virus (RSV) infection and COVID-19. RSV infection remains one of the leading causes of hospitalisation in infants [71], and there are no approved intranasal or inhaled RSV vaccines at the time of writing. Research on RSV vaccines has also been motivated by the recent surge in RSV incidence in various parts of the world [72–74]. Social restrictions during the COVID-19 pandemic have not only limited the spread of COVID-19 but also other respiratory infections, including RSV, which resulted in a cohort of children with no natural immunity against RSV [72]. The relaxation of these restrictions has led to a spike in RSV infections in children during atypical season. Some promising results on recent studies of RSV vaccines have been reported. For instance, RSV could be attenuated through deletion of the RNA synthesis regulatory protein M2-2, which led to an increase in gene transcription and antigen expression, but a reduction in genome replication. This vaccine candidate was administered in RSV-seronegative children of age 6–24 months as a single intranasal dose. At least 85% of vaccinated individuals demonstrated a fourfold increase or more in serum neutralising antibodies [75–77]. In another study, live-attenuated RSV lacking the G attachment protein (RSVΔG), which involves viral binding to host cells, was administered intranasally to healthy adults of age from 18 to 50 years [78]. The vaccine was well-tolerated, but the extent of systemic immunogenicity was limited, attributed in part to the fact that this cohort was already seropositive at baseline before vaccination.

Despite the initial success of the injectable COVID-19 vaccinations, enormous work has been dedicated to the development of mucosal vaccines against SARS-CoV-2 to address the lack of mucosal immunity induction in injectable vaccines. To illustrate, an intranasal subunit vaccine based on freeze-dried spike protein formulated with a liposomal stimulator of interferon gene (STING) agonist as an adjuvant has been reported, with the goal to improve immunisation against SARS-CoV-2 [79]. BALB/c mice receiving a single dose of intranasal vaccine-induced IgA secretion in both the nasal cavity and the lung, as well as serum anti-spike IgG. Intranasal administration of ChAdOx1 nCoV-19 vaccines (Oxford-AstraZeneca COVID-19 vaccine) was demonstrated to reduce viral shedding upon challenge in both hamster and rhesus macaque animal models [80]. Syrian hamsters were protected from pulmonary damage when challenged by the alpha or beta variants [81], and potent systemic and mucosal immunity was induced in the murine model [82]. Contrary to these promising results in preclinical studies, intranasal vaccination with ChAdOx1 nCoV-19 in healthy adults was shown to be safe yet neither a profound systemic response nor consistent mucosal immunity has been detected [83]. Adenovirus type-5 vector-based COVID-19 vaccines represent another prospective candidate for mucosal vaccination. Preclinical studies reported that macaques were protected against the SARS-CoV-2 challenge after receiving a single intranasal Ad5-nCoV vaccination [84], and profound systemic and mucosal immune responses were elicited when the vaccine was administered by nebulisation [85]. Similar immunogenicity was also observed in mice upon intranasal vaccination of Ad5.SARS-CoV-2-S1 [86]. Preliminary report on a phase I clinical trial on nebulised Ad5-nCoV vaccine showed that the vaccine was generally well-tolerated, and two doses of nebulised vaccines 28 days apart could induce a humoral immune response similar to one dose of intramuscular injection [87]. Remarkably, intranasal immunisation as a booster dose to a primary intramuscular vaccination strongly enhanced humoral immune responses. In a separate study involving more than 700 participants naïve to SARS-CoV-2 vaccination and infection, the vaccine candidate dNS1-RBD was given in two intranasal doses 14 or 21 days apart [88]. dNS1-RBD was generally well-tolerated, yet only a modest mucosal and humoral immune response was recorded, with seroconversion observed and local RBD-specific sIgA detected in 10–22% and 12–13% of vaccine recipients, respectively. The landscape of intranasal and inhaled vaccines against SARS-CoV-2 is rapidly evolving. Readers interested in a more thorough account of their latest development are referred to review articles recently published [89–93].

3 Social and Economic Benefits of Intranasal and Inhaled Vaccines

Apart from the broader protection provided through both mucosal and systemic immune responses, inhaled and intranasal vaccines offer several additional benefits over their injectable counterparts. Self-administration using inhaler devices or nasal sprays is possible for most healthy individuals. Training through videos [94, 95] or mobile devices [96, 97] can be provided when social distancing is exercised or access to health care services is limited. This clearly juxtaposes injectables that require close contact with trained personnel for vaccination. The route of administration is relevant especially when the safe use of injectable vaccines relies on the correct injection technique. To exemplify, the risks of acute myocarditis or myopericarditis associated with mRNA vaccines against COVID-19 [98, 99] may further increase when they are inadvertently given intravenously instead of the intended intramuscular injection, as evident in an in vivo study using murine model [100]. Similarly, risks of thrombotic thrombocytopenia may increase in vaccines based on adenoviral vectors when accidentally administered intravenously [101]. With respect to this, mucosal vaccines are safer as they are locally administered with better tolerance on the site of antigen deposition or expression and have a lower risk of administration error. Inhaled or intranasal vaccines do not require sterile drug reconstitution and administration, thus mass vaccination programmes can be carried out in less stringent settings. Healthcare professionals and resources, which could be in huge demand during a pandemic, can be spared and allocated to patients who require greater medical attention.

Intranasal and inhaled vaccines also eliminate the need for needles for injection, which could be appealing to people with trypanophobia (needle phobia) and potentially address the associated vaccine hesitancy. The prevalence of trypanophobia was estimated to lie between 20–50% in adolescents and 20–30% in young adults [102]. While needle fear is generally less prominent with increasing age, it still constitutes to approximately 8–27% of avoidance of influenza vaccination among hospital workers and adult patients. Similar figures have been reported by another survey in the UK [103]. As much as 26.2% of the adult population was screened positive for blood-injection-injury phobia, which might account for about 10% of cases of COVID-19 vaccine hesitancy. Non-invasive administration also eliminates the risk of bloodborne disease transmission due to cross-contamination or needle-stick injury, which could be more relevant in less-developed regions where health-care services are less accessible. Recent meta-analyses reported that the annual global prevalence of needle-stick injuries in healthcare workers was around 44%, with that in the Southeast Asia region and Africa reaching above 50% [104, 105]. Inhaler devices or nasal sprays pose less risk of cross-contamination. They can be disinfected more readily in a household setting with 70% ethanol or diluted sodium hypochlorite if needed for reuse or before disposal as domestic waste [106, 107]. The shear demand for inhaler devices may also motivate and accelerate the development of inhaler devices made of sustainable materials [108], as well as single-use disposable inhalers [109, 110]. On the other hand, used needles must be handled as clinical sharps, adding extra costs to waste management and putting additional risks to the environment [111]. This poses a substantial burden particularly among countries with limited resources that still heavily exercise suboptimal waste disposal protocols such as opening and burning syringes and sharps without burying them, disposing glass vials without disinfection, or direct dumping, thus jeopardising both the human and environmental health [112].

All vaccines currently on the market are administered as liquids, including those formulated in solid dosage forms that require reconstitution prior to administration. Regardless of the platform, vaccines are primarily biopharmaceuticals with poor stability in liquid formulations that often require cold-chain logistics. This increases the cost of vaccine distribution and hampers vaccine coverage, which can lead to considerable public health impacts in a race against time during a pandemic. The short shelf-life combined with the requirement of strict storage conditions implies that stockpiling vaccines in preparation for a disease outbreak can be expensive to maintain. These shortcomings have driven research of formulating biopharmaceuticals including vaccines into solid dosage forms for improved stability [113–116]. By manufacturing vaccines as dry powder for reconstitution prior to administration, the cost associated with vaccine production, transportation, storage, and wastage because of the short shelf-life of liquid formulations can be reduced [117]. Nonetheless, only part of the problems can be addressed since additional manpower and resources must be allocated to vaccine reconstitution, which can be an error-prone procedure. Vaccine-related errors such as inadequate mixing, inaccurate aspiration of reconstitution vials, and spillage or leakage of vaccines during handling are not uncommon [118], especially for multi-component formulations in which vaccine components are physically stored in separate containers [119]. Patients could inadvertently receive diluent only [120] or undiluted vaccine concentrate [121]. These risks could be avoided in intranasal and inhaled vaccines which permit direct administration of vaccines.

4 Challenges and Limitations of Intranasal and Inhaled Vaccines

Despite the unique merits offered by intranasal and inhaled vaccines, their formulation development presents a distinctive set of challenges that require further research, as elaborated below.

4.1 Vaccine Adjuvants and Formulation Design

Immunisation at the mucosae is a double-edged sword. On the one hand, mucosal immunity at the entry point of pathogens offers early neutralisation and host immune responses against infection, even with the potential of achieving sterilising immunity [122, 123]. Strong mucosal immunity is effective in reducing infection and halting disease transmission, thereby contributing to herd immunity. On the other hand, as the mucosa is exposed to the environment, it is evolved to tolerate external stimuli to prevent immune responses or inflammation from being constitutively activated, which could otherwise be overwhelming [124, 125]. Therefore, intranasal or inhaled vaccines must be engineered to elicit adequate immune response effectively. Traditional platforms like live-attenuated or inactivated vaccines are usually preferred for their stronger immunogenicity, including FluMist® Quadrivalent, the first intranasal vaccine in the market. Conversely, emerging vaccine technologies, namely, subunit- or nucleic acid-based vaccines, often suffer from their reduced immunogenicity in exchange for improved safety, especially when administered locally. The problem of insufficient immune response is complicated by the limited availability of effective adjuvants. Adjuvants are excipients often included in vaccine formulations to enhance immunogenicity. Typical adjuvants such as aluminium salts (e.g., hydrated potassium aluminium sulphate, or alum) are used in injectable vaccines. Their mechanism as adjuvants remains poorly understood and might be relevant to the route of administration [126], with hypotheses including acting as a depot on which the antigens adsorb for a sustained release resulting in a long-lasting immune response, and the induction of strong innate immune responses following mild inflammation at the injection site [127, 128]. The role of aluminium-based adjuvants for intranasal vaccines has been inconsistent [129]. Other adjuvants for mucosal vaccines are being investigated, such as silica nanoparticles (SiO_2), cyclic-dimeric guanosine monophosphate (c-di-GMP) [43, 130] and immune potentiators [131] like polyinosinic:polycytidylic acid (Poly(I:C)). The immunogenicity of intranasal vaccines has been shown to be heavily influenced by the adjuvants used [132], yet the full mechanism of immunopotentiation of these adjuvants remains to be elucidated.

In addition to the potentially complex and delicate molecular structures of vaccines, especially the novel vaccine platforms, the task of designing vaccine formulation is further challenged by the lack of approved excipients for inhalation, especially for dry powder formulations. Formulation excipients should exhibit desirable critical attributes, such as a good safety profile, compatibility with the vaccine components and adjuvants, and the capacity to confer formulation stability during manufacture, storage, and aerosolisation [133]. pH buffers, osmolarity regulators, and preservatives are general excipients used in liquid formulations of biopharmaceuticals including vaccines. Excipients commonly used in the development of dry powder vaccine formulations include drying stabilisers such as mannitol, trehalose, lactose, and dextran [133–135]. For inhaled particle formulations, dispersion enhancers can be incorporated to improve aerosol performance, with L-leucine being the most investigated excipient [136, 137]. On the other hand, intranasal formulations contain functional excipients to increase drug retention time and reduce the rate of mucociliary clearance, by using mucoadhesive agents such as chitosan and cellulose derivatives [138, 139]. Microcrystalline cellulose or hypromellose also serves as a viscosity agent to prolong the residence time of liquid formulations in the nasal cavity [140]. Given that vaccines are biomacromolecules that are prone to degradation during atomisation and drying, early incorporation of suitable excipients in the formulation design is indispensable to the success of an intranasal or inhalable vaccine.

4.2 Dose Variability and Administration Device

Compared to injections, intranasal administration and inhalation could be subject to larger dose variability. In intranasal administration, the vaccine is actively dispersed into the nostril as liquid or powder aerosol using a nasal spray. While the emitted vaccine dose can be controlled, the amount of antigen that deposits in the nasal cavity depends on several factors [141], such as the surface tension and viscosity of the liquid formulation, as well as the spray plume angles, droplet sizes and spray velocities of the spray device. These factors can be optimised during formulation development [142]. It may be desirable for intranasal vaccines to be deposited at the nasal-associated lymphoid tissue (NALT) located at the nasopharynx where it houses an abundant number of APCs and lymphocytes [143]. On the contrary, patient factors are less controllable, such as individual variations in anatomy and pathophysiology of the nasal cavity, including the efficiency of mucociliary clearance [144, 145], as well as the variations in the use of devices, including the insertion depth of the device and the administration angle [146]. They also exert a profound effect on deposited dose and constitute dose variability.

With respect to inhalation, dose variability can be even more pronounced, and each type of inhaler device has its own strengths and constraints in delivering vaccines (Table 4). Metered-dose inhalers (MDIs) and Soft Mist™ Inhalers (SMI) [147] are less relevant for the delivery of vaccines as they have limited drug payload and are better suited for delivering potent small molecules. Nebulisers are frequently used in clinical trials for the delivery of aerosolised vaccine candidates [87, 148–150]. They offer distinct advantages. Firstly, investigational vaccines can be formulated into simple aqueous solutions and be aerosolised using existing nebulisers. It eliminates certain formulation challenges associated with other inhaler devices, such as biocompatibility with propellant in MDIs, or particle engineering in dry powder inhalers (DPIs), thereby significantly reducing the barriers of formulation at the early phase of new vaccine development. The drug reservoir design also permits a high dose of vaccines to be delivered. Since an external energy source is used to generate slow-moving respirable droplets, vaccine aerosols can be inhaled during tidal breathing without the need for hand-breath coordination. This makes nebulisation more suitable for people with limited inspiration capacity. However, nebulisation also comes with its downsides. The administration time is long, which would become a bottleneck to the inoculation rate during a mass vaccination program. Nebulisers are also associated with the risk of potential cross-contamination to the environment by droplets aspirated from infected patients [151]. It can be particularly concerning during an active outbreak of an epidemic or pandemic. The risk of contamination can be reduced through adequate ventilation and the use of outlet filters, yet these measures inevitably lead to additional costs to immunisation. A dry powder inhaler presents as an attractive device for the delivery of inhaled vaccines. It is the only kind of device that allows direct pulmonary administration of vaccines in dry powder formulation, thus offering superior product stability and reduced logistic cost. Most DPIs are passive devices that rely solely on the inspiration force

Table 4 Strengths and limitations of different inhaler devices with respect to the delivery of vaccines

Devices	Strengths	Limitations
Nebulisers	Simple formulation design; 'go-to' device in clinical trials for vaccine candidates	Administration time can be long (minutes)
	Less compatibility issues	Potential contamination of the surrounding environment by exhaled droplets [151]
	Virtually unlimited payload.	
	Suit most patients	
Metered-dose inhalers	Low cost	Compatibility issues with organic propellants
	Disposable device [110]	High shear stress during aerosolisation
		The payload can be limited
Soft mist™ inhaler	Improved deep lung delivery over MDI	The payload can be limited
		Proprietary device [147]
Dry powder inhalers	Higher drug payload (tens of milligrams) [157]	Formulation design and particle engineering can be challenging
	Direct administration of solid formulations	May not suit patients with insufficient inspiratory force
	Unit-dose design available	
	Longer shelf-life	
	Disposable device	

of the patients for powder dispersion and aerosolisation. While there have been many clinical trials on intranasal and inhaled vaccines of liquid formulations, there is a very limited clinical trial of inhalable dry powder vaccine, with one formulation targeting measles that was prepared by supercritical fluid drying [60, 152]. Even with optimised physicochemical and aerosol properties of the formulation, difference in effective deposited dose remains inevitable because of individual variations such as lung functions and device manoeuvre technique [153–155]. DPIs also possess a limit on the powder dose that can be delivered [156, 157], and devices capable of delivering high powder doses are under active research [158].

Notwithstanding the various devices available for drug delivery through intranasal route or inhalation, there are several common considerations that deserve more attention before their adaptation for vaccination. Notably, immune response and safety of intranasal and inhaled vaccines must be meticulously evaluated in clinical trials to ensure proper dosage, given their inherently larger dose variability. Dose determination could be further complicated by the aforementioned variations in mucosal tolerance between individuals. Intranasal or inhaled vaccines can also result in a less consistent particle deposition along the respiratory tract. For example, a fraction of particles may deposit at the back of the throat and be swallowed [159]. The evaluation of the effects of such off-target deposition on immune response and vaccine safety becomes crucial. The less predictable deposition pattern could result in different pharmacokinetic and pharmacodynamic properties with more pronounced variations that are distinct from their injectable counterparts. It is prudent to evaluate not only the extent and the duration of the triggered immune responses such as serum immunoglobulin levels, but also the pharmacokinetics of the vaccine particles, especially with the newer vaccine platforms. Regarding vaccine delivery, devices like MDI and nebuliser are designed to store and deliver multiple doses of medications for managing chronic respiratory disorders, whereas unit dose devices are more common for nasal spray and DPI. Provided that vaccination is only given as a limited number of doses separated by weeks or months, portable and disposable single-use device that can be manufactured at large scale and at low cost is suited for such purpose [109, 110].

4.3 Preclinical and Clinical Studies, and Inadequate Public Recognition

Preclinical studies of intranasal and inhaled vaccines can be difficult because they require appropriate animal models. Ideally, they should exhibit relevant immunogenicity to the vaccine and adjuvants with comparable pathophysiology to the targeted pathogens, while permitting vaccines to be given by the intended route of administration. Studies involving multiple animal models are not uncommon. The lack of in vitro-in vivo correlation for inhalable formulations also implies that clinical trials of vaccines are often needed [160]. Although healthy adults are often recruited in early phases of clinical trials for their greater recovery capacity in case of adverse reactions, thorough and careful evaluations must be conducted when the use of investigational vaccines is extended to susceptible populations such as the elderly, children and infants. Their pulmonary functions and airway anatomy can differ from healthy adults, which could affect the delivered dose and dose variability of intranasal and inhaled vaccines. Additionally, as these cohorts are usually more vulnerable to infections and thus benefit more from vaccinations, they are prioritised to be immunised. Depending on the targeted diseases, healthy adults could already be seropositive at the baseline [78], confounding the interpretation of the outcomes.

Finally, recognising the significance of inhaled or intranasal vaccines as a viable or even superior alternative to conventional injectable vaccines by healthcare professionals and policymakers is of paramount importance to the development of future intranasal and inhaled vaccines. The utility of intranasal and inhaled vaccines has often been ignored or disguised by their slower response to an emerging epidemic or pandemic. With respect to inhaled vaccines, the inhaler devices must also be incorporated during product development. To accelerate the process, continuous research is important even when the epidemic is over. Moreover, mRNA vaccines offer higher modularity in formulation development as the expression of antigens is depending on the mRNA sequence [161]. Thus, existing formulations can be promptly adopted to deliver different sequences of mRNA, as illustrated in the bivalent COVID-19 vaccines expressing spike proteins of both the wild-type SARS-CoV-2 and the omicron variant [162]. Only with sufficient understanding can the longer development lead time be justified by the many inimitable benefits of intranasal and inhaled vaccines, specifically their unique potential in inducing mucosal immunity to achieve prevention of infection or even termination of disease transmission.

4.4 Vaccine Manufacture

Large-scale production of vaccines is paramount to effective vaccine dissemination, the challenge of which depends on different factors, including the vaccine platform, vaccine dosage formulation, as well as the route of administration. Among the three marketed intranasal vaccines (Table 2), the live-attenuated vaccines FluMist® are produced using eggs, which are time-consuming and labour-intensive [163]. The long lead time implies that they can be less responsive to off-season outbreaks of communicable diseases. The scale of production can also be limited by the availability of eggs. A possible answer to this manufacturing challenge is the use of newer vaccine platforms. As illustrated by the other two recently marketed COVID-19 viral vector vaccines, Convidecia® Air and iNCOVACC®, which are both manufactured with recombinant technologies using human cells as the host of virus replication [122, 164], these newer vaccine platforms not only offer improved safety profile and fewer infection risks but they can also be manufactured more rapidly and scaled up more easily, without using animal derived raw materials or even completely cell-free (for mRNA vaccines) [165]. The manufacturing potential brought by these new vaccine platforms has been convincingly demonstrated by the COVID-19 mRNA vaccines that have been developed, produced and distributed at a pace so fast and scale so large that has never been achieved [166].

The dosage formulation of vaccines also becomes relevant for intranasal and inhaled vaccines because these routes permit direct administration of either liquid or solid preparations. While liquid formulations can be manufactured similarly as injectable vaccines, it is more challenging to prepare solid dosage formulations of vaccines suitable for intranasal administration or inhalation. Lyophilisation (freeze-drying) is commonly utilised for the manufacture of solid-form biopharmaceuticals. Although the resultant cake of freeze-dried pharmaceuticals can be readily reconstituted into solutions prior to administration, as with many injectable vaccines, it is not suitable for direct inhalation or intranasal administration because it lacks appropriate particle size distribution and is difficult to disperse. Variations of lyophilisation, such as spray-freeze drying, have been used to produce dry formulations that are suitable for intranasal or inhaled administration [167]. Nonetheless, these techniques are less scalable and the use of spray-freeze drying at an industrial scale with high throughput is undergoing active research [168]. Spray drying is a popular particle engineering technique that has been used to produce solid biopharmaceutical formulations capable of direct intranasal administration or inhalation, including vaccines [169]. It combines size reduction and evaporation into one continuous process and can be scaled up, rendering it a promising approach for the manufacturing of solid intranasal and inhaled vaccines.

5 Summary

With the unique advantages of intranasal and inhaled vaccines, particularly the potential to confer sterile immunity against airborne or droplet-transmitted communicable respiratory diseases through the induction of mucosal immunity, the prospect of intranasal and inhaled vaccines is certainly promising and justifies continual research to overcome the development challenges. Future research directions include improving formulation stability and prolonging shelf-life to facilitate mass vaccine dissemination, as well as identifying safe and potent vaccine adjuvants to induce robust mucosal immune responses. Since the various vaccine platforms are fundamentally biopharmaceuticals of different classes, it is anticipated that the development of intranasal and inhaled vaccines could be expedited by the advancement of intranasal and inhaled formulations of other biopharmaceuticals, as outlined in other chapters of this book. To achieve optimal protection, the therapeutic position of intranasal and inhaled vaccines, either as primary vaccination or as booster doses to primary injectable vaccines, should also be explored [170]. The success of introducing new vaccine platforms would allow us to be better equipped in our arsenal to fight against existing as well as emerging respiratory infections.

References

1. Ong CWM, et al. Epidemic and pandemic viral infections: impact on tuberculosis and the lung: a consensus by the World Association for Infectious Diseases and Immunological Disorders (WAidid), Global Tuberculosis Network (GTN), and members of the European Society of Clinical Microbiology and Infectious Diseases Study Group for Mycobacterial Infections (ESGMYC). Eur Respir J. 2020;56(4)
2. Shapshak P, et al. The influenza pandemic of 2009: lessons and implications. Mol Diagn Ther. 2011;15(2):63–81.
3. Li Q, et al. Early transmission dynamics in Wuhan, China, of novel coronavirus-infected pneumonia. N Engl J Med. 2020;382(13):1199–207.
4. Zhu N, et al. A novel coronavirus from patients with pneumonia in China, 2019. N Engl J Med. 2020;382(8):727–33.
5. Hilleman MR. Vaccines in historic evolution and perspective: a narrative of vaccine discoveries. Vaccine. 2000;18(15):1436–47.
6. Plotkin SA, Plotkin SL. The development of vaccines: how the past led to the future. Nat Rev Microbiol. 2011;9(12):889–93.
7. Zepp F. Principles of vaccine design-lessons from nature. Vaccine. 2010;28(Suppl 3):C14–24.
8. Ghattas M, et al. Vaccine technologies and platforms for infectious diseases: current Progress, challenges, and opportunities. Vaccines (Basel). 2021;9(12):1490.
9. Rodriguez L, et al. Comparative study of the temperature sensitive, cold adapted and attenuated mutations present in the master donor viruses of the two commercial human live attenuated influenza vaccines. Viruses. 2019;11(10):928.
10. Khoshnood S, et al. An overview on inactivated and live-attenuated SARS-CoV-2 vaccines. J Clin Lab Anal. 2022;36(5):e24418.
11. Handabile C, et al. Inactivated whole virus particle influenza vaccine induces anti-neuraminidase antibodies that may contribute to cross-protection against heterologous virus infection. Vaccines (Basel). 2022;10(5):804.

12. Ho JK, Jeevan-Raj B, Netter HJ. Hepatitis B Virus (HBV) subviral particles as protective vaccines and vaccine platforms. Viruses. 2020;12(2):126.
13. Parums DV. Editorial: first approval of the protein-based Adjuvanted Nuvaxovid (NVX-CoV2373) Novavax vaccine for SARS-CoV-2 could increase vaccine uptake and provide immune protection from viral variants. Med Sci Monit. 2022;28:e936523.
14. Gebre MS, et al. Novel approaches for vaccine development. Cell. 2021;184(6):1589–603.
15. Jacob-Dolan C, Barouch DH. COVID-19 vaccines: adenoviral vectors. Annu Rev Med. 2022;73:41–54.
16. Szabo GT, Mahiny AJ, Vlatkovic I. COVID-19 mRNA vaccines: platforms and current developments. Mol Ther. 2022;30(5):1850–68.
17. Pushparajah D, et al. Advances in gene-based vaccine platforms to address the COVID-19 pandemic. Adv Drug Deliv Rev. 2021;170:113–41.
18. Lavelle EC, Ward RW. Mucosal vaccines – fortifying the frontiers. Nat Rev Immunol. 2022;22(4):236–50.
19. Miquel-Clopes A, et al. Mucosal vaccines and technology. Clin Exp Immunol. 2019;196(2):205–14.
20. Holmgren J, Czerkinsky C. Mucosal immunity and vaccines. Nat Med. 2005;11(4 Suppl):S45–53.
21. Azzi L, et al. Mucosal immune response in BNT162b2 COVID-19 vaccine recipients. EBioMedicine. 2022;75:103788.
22. Russell MW, et al. Mucosal immunity in COVID-19: a neglected but critical aspect of SARS-CoV-2 infection. Front Immunol. 2020;11:611337.
23. Roh EH, Fromen CA, Sullivan MO. Inhalable mRNA vaccines for respiratory diseases: a roadmap. Curr Opin Biotechnol. 2022;74:104–9.
24. Shah SM, et al. A review on current COVID-19 vaccines and evaluation of particulate vaccine delivery systems. Vaccines (Basel). 2021;9(10):1086.
25. Tschernig T, Pabst R. Bronchus-associated lymphoid tissue (BALT) is not present in the normal adult lung but in different diseases. Pathobiology. 2000;68(1):1–8.
26. Tan HX, et al. Inducible bronchus-associated lymphoid tissues (iBALT) serve as sites of B cell selection and maturation following influenza infection in mice. Front Immunol. 2019;10:611.
27. Neyt K, et al. Early IL-1 Signaling promotes iBALT induction after influenza virus infection. Front Immunol. 2016;7:312.
28. John-Schuster G, et al. Cigarette smoke-induced iBALT mediates macrophage activation in a B cell-dependent manner in COPD. Am J Physiol Lung Cell Mol Physiol. 2014;307(9):L692–706.
29. Hiramatsu K, et al. Inhalation of diesel exhaust for three months affects major cytokine expression and induces bronchus-associated lymphoid tissue formation in murine lungs. Exp Lung Res. 2003;29(8):607–22.
30. Rangel-Moreno J, et al. Inducible bronchus-associated lymphoid tissue (iBALT) in patients with pulmonary complications of rheumatoid arthritis. J Clin Invest. 2006;116(12):3183–94.
31. Silva-Sanchez A, Randall TD. Role of iBALT in respiratory immunity. Curr Top Microbiol Immunol. 2020;426:21–43.
32. Fleige H, et al. IL-17-induced CXCL12 recruits B cells and induces follicle formation in BALT in the absence of differentiated FDCs. J Exp Med. 2014;211(4):643–51.
33. Kimura S, et al. Airway M cells arise in the lower airway due to RANKL signaling and reside in the bronchiolar epithelium associated with iBALT in murine models of respiratory disease. Front Immunol. 2019;10:1323.
34. Halle S, et al. Induced bronchus-associated lymphoid tissue serves as a general priming site for T cells and is maintained by dendritic cells. J Exp Med. 2009;206(12):2593–601.
35. Oh JE, et al. Intranasal priming induces local lung-resident B cell populations that secrete protective mucosal antiviral IgA. Sci Immunol. 2021;6(66):eabj5129.
36. Woof JM, Kerr MA. The function of immunoglobulin A in immunity. J Pathol. 2006;208(2):270–82.

37. Corthesy B. Multi-faceted functions of secretory IgA at mucosal surfaces. Front Immunol. 2013;4:185.
38. Kato A, et al. B-lymphocyte lineage cells and the respiratory system. J Allergy Clin Immunol. 2013;131(4):933–57.
39. Bakema JE, van Egmond M. The human immunoglobulin A Fc receptor FcalphaRI: a multi-faceted regulator of mucosal immunity. Mucosal Immunol. 2011;4(6):612–24.
40. Balu S, et al. A novel human IgA monoclonal antibody protects against tuberculosis. J Immunol. 2011;186(5):3113–9.
41. Tran AC, et al. Mucosal therapy of multi-drug resistant tuberculosis with IgA and interferon-gamma. Front Immunol. 2020;11:582833.
42. Koernig S, et al. Topical application of human-derived Ig isotypes for the control of acute respiratory infection evaluated in a human CD89-expressing mouse model. Mucosal Immunol. 2019;12(4):1013–24.
43. Neuhaus V, et al. A new adjuvanted nanoparticle-based H1N1 influenza vaccine induced antigen-specific local mucosal and systemic immune responses after administration into the lung. Vaccine. 2014;32(26):3216–22.
44. Tamura S, et al. Intranasal inactivated influenza vaccines: a reasonable approach to improve the efficacy of influenza vaccine? Jpn J Infect Dis. 2016;69(3):165–79.
45. Qin M, Du G, Sun X. Recent advances in the noninvasive delivery of mRNA. Acc Chem Res. 2021;54(23):4262–71.
46. Cox RJ, Brokstad KA, Ogra P. Influenza virus: immunity and vaccination strategies. Comparison of the immune response to inactivated and live, attenuated influenza vaccines. Scand J Immunol. 2004;59(1):1–15.
47. Alu A, et al. Intranasal COVID-19 vaccines: from bench to bed. EBioMedicine. 2022;76:103841.
48. Suryadevara N, et al. A molecular signature of lung-resident CD8(+) T cells elicited by subunit vaccination. Sci Rep. 2022;12(1):19101.
49. Son YM, et al. Tissue-resident CD4(+) T helper cells assist the development of protective respiratory B and CD8(+) T cell memory responses. Sci Immunol. 2021;6(55):eabb6852.
50. Nizard M, et al. Induction of resident memory T cells enhances the efficacy of cancer vaccine. Nat Commun. 2017;8:15221.
51. Slutter B, et al. Dynamics of influenza-induced lung-resident memory T cells underlie waning heterosubtypic immunity. Sci Immunol. 2017;2(7):eaag2031.
52. Reagin KL, Klonowski KD. Incomplete memories: the natural suppression of tissue-resident memory CD8 T cells in the lung. Front Immunol. 2018;9:17.
53. Swarnalekha N, et al. T resident helper cells promote humoral responses in the lung. Sci Immunol. 2021;6(55):eabb6808.
54. Bienenstock J, et al. A common mucosal immunologic system involving the bronchus, breast and bowel. Adv Exp Med Biol. 1978;107:53–9.
55. Tulic MK, Piche T, Verhasselt V. Lung-gut cross-talk: evidence, mechanisms and implications for the mucosal inflammatory diseases. Clin Exp Allergy. 2016;46(4):519–28.
56. Bhide Y, et al. Pulmonary delivery of influenza vaccine formulations in cotton rats: site of deposition plays a minor role in the protective efficacy against clinical isolate of H1N1pdm virus. Drug Deliv. 2018;25(1):533–45.
57. Tomar J, et al. Pulmonary immunization: deposition site is of minor relevance for influenza vaccination but deep lung deposition is crucial for hepatitis B vaccination. Acta Pharm Sin B. 2019;9(6):1231–40.
58. Minne A, et al. The delivery site of a monovalent influenza vaccine within the respiratory tract impacts on the immune response. Immunology. 2007;122(3):316–25.
59. Jeyanathan V, et al. Differential biodistribution of adenoviral-vectored vaccine following intranasal and endotracheal deliveries leads to different immune outcomes. Front Immunol. 2022;13:860399.

60. Heida R, Hinrichs WL, Frijlink HW. Inhaled vaccine delivery in the combat against respiratory viruses: a 2021 overview of recent developments and implications for COVID-19. Expert Rev Vaccines. 2022;21(7):957–74.
61. de Swart RL, et al. Needle-free delivery of measles virus vaccine to the lower respiratory tract of non-human primates elicits optimal immunity and protection. NPJ Vaccines. 2017;2:22.
62. Meyer M, et al. Aerosolized Ebola vaccine protects primates and elicits lung-resident T cell responses. J Clin Invest. 2015;125(8):3241–55.
63. Woo SJ, et al. Intranasal immunization with protective antigen of Bacillus anthracis induces a long-term immunological memory response. Mol Immunol. 2015;67(2 Pt B):492–500.
64. Sou T, et al. Spray-dried influenza antigen with Trehalose and leucine produces an Aerosolizable powder vaccine formulation that induces strong systemic and mucosal immunity after pulmonary administration. J Aerosol Med Pulm Drug Deliv. 2015;28(5):361–71.
65. Ainai A, et al. Intranasal vaccination with an inactivated whole influenza virus vaccine induces strong antibody responses in serum and nasal mucus of healthy adults. Hum Vaccin Immunother. 2013;9(9):1962–70.
66. Mohn KG, et al. Immune responses after live attenuated influenza vaccination. Hum Vaccin Immunother. 2018;14(3):571–8.
67. Dibben O, et al. Defining the root cause of reduced H1N1 live attenuated influenza vaccine effectiveness: low viral fitness leads to inter-strain competition. NPJ Vaccines. 2021;6(1):35.
68. Eiden J, et al. Intranasal M2SR (M2-deficient single replication) H3N2 influenza vaccine provides enhanced mucosal and serum antibodies in adults. J Infect Dis. 2022;227(1):103–12.
69. Nicolodi C, et al. Safety and immunogenicity of a replication-deficient H5N1 influenza virus vaccine lacking NS1. Vaccine. 2019;37(28):3722–9.
70. Pan SC, et al. A double-blind, randomized controlled trial to evaluate the safety and immunogenicity of an intranasally administered trivalent inactivated influenza vaccine with the adjuvant LTh(alphaK): a phase II study. Vaccine. 2020;38(5):1048–56.
71. Wildenbeest JG, et al. The burden of respiratory syncytial virus in healthy term-born infants in Europe: a prospective birth cohort study. Lancet Respir Med. 2023;11(4):341–53.
72. Bardsley M, et al. Epidemiology of respiratory syncytial virus in children younger than 5 years in England during the COVID-19 pandemic, measured by laboratory, clinical, and syndromic surveillance: a retrospective observational study. Lancet Infect Dis. 2023;23(1):56–66.
73. Kuitunen I, et al. Respiratory virus circulation in children after relaxation of COVID-19 restrictions in fall 2021 – a nationwide register study in Finland. J Med Virol. 2022;94(9):4528–32.
74. Sim JY, et al. Circulating pediatric respiratory pathogens in Taiwan during 2020: dynamic change under low COVID-19 incidence. J Microbiol Immunol Infect. 2022;55(6 Pt 2):1151–8.
75. McFarland EJ, et al. Live-attenuated respiratory syncytial virus vaccine candidate with deletion of RNA synthesis regulatory protein M2-2 is highly immunogenic in children. J Infect Dis. 2018;217(9):1347–55.
76. McFarland EJ, et al. Live respiratory syncytial virus attenuated by M2-2 deletion and stabilized temperature sensitivity mutation 1030s is a promising vaccine candidate in children. J Infect Dis. 2020;221(4):534–43.
77. McFarland EJ, et al. Live-attenuated respiratory syncytial virus vaccine with M2-2 deletion and with small hydrophobic noncoding region is highly immunogenic in children. J Infect Dis. 2020;221(12):2050–9.
78. Verdijk P, et al. First-in-human administration of a live-attenuated RSV vaccine lacking the G-protein assessing safety, tolerability, shedding and immunogenicity: a randomized controlled trial. Vaccine. 2020;38(39):6088–95.
79. An X, et al. Single-dose intranasal vaccination elicits systemic and mucosal immunity against SARS-CoV-2. iScience. 2021;24(9):103037.
80. van Doremalen N, et al. Intranasal ChAdOx1 nCoV-19/AZD1222 vaccination reduces viral shedding after SARS-CoV-2 D614G challenge in preclinical models. Sci Transl Med. 2021;13(607):eabh0755.

81. Fischer RJ, et al. ChAdOx1 nCoV-19 (AZD1222) protects Syrian hamsters against SARS-CoV-2 B.1.351 and B.1.1.7. Nat Commun. 2021;12(1):5868.
82. Cokaric Brdovcak M, et al. ChAdOx1-S adenoviral vector vaccine applied intranasally elicits superior mucosal immunity compared to the intramuscular route of vaccination. Eur J Immunol. 2022;52(6):936–45.
83. Madhavan M, et al. Tolerability and immunogenicity of an intranasally-administered adenovirus-vectored COVID-19 vaccine: an open-label partially-randomised ascending dose phase I trial. EBioMedicine. 2022;85:104298.
84. Feng L, et al. An adenovirus-vectored COVID-19 vaccine confers protection from SARS-COV-2 challenge in rhesus macaques. Nat Commun. 2020;11(1):4207.
85. Xu F, et al. Safety, mucosal and systemic immunopotency of an aerosolized adenovirus-vectored vaccine against SARS-CoV-2 in rhesus macaques. Emerg Microbes Infect. 2022;11(1):438–41.
86. Kim E, et al. A single subcutaneous or intranasal immunization with adenovirus-based SARS-CoV-2 vaccine induces robust humoral and cellular immune responses in mice. Eur J Immunol. 2021;51(7):1774–84.
87. Wu S, et al. Safety, tolerability, and immunogenicity of an aerosolised adenovirus type-5 vector-based COVID-19 vaccine (Ad5-nCoV) in adults: preliminary report of an open-label and randomised phase 1 clinical trial. Lancet Infect Dis. 2021;21(12):1654–64.
88. Zhu F, et al. Safety and immunogenicity of a live-attenuated influenza virus vector-based intranasal SARS-CoV-2 vaccine in adults: randomised, double-blind, placebo-controlled, phase 1 and 2 trials. Lancet Respir Med. 2022;10(8):749–60.
89. Kar S, et al. Oral and intranasal vaccines against SARS-CoV-2: current progress, prospects, advantages, and challenges. Immun Inflamm Dis. 2022;10(4):e604.
90. Dhama K, et al. COVID-19 intranasal vaccines: current progress, advantages, prospects, and challenges. Hum Vaccin Immunother. 2022;18(5):2045853.
91. Miteva D, et al. Mucosal COVID-19 vaccines: risks, benefits and control of the pandemic. World J Virol. 2022;11(5):221–36.
92. Chavda VP, et al. Intranasal vaccines for SARS-CoV-2: from challenges to potential in COVID-19 management. Drug Discov Today. 2021;26(11):2619–36.
93. Focosi D, Maggi F, Casadevall A. Mucosal vaccines, sterilizing immunity, and the future of SARS-CoV-2 virulence. Viruses. 2022;14(2):187.
94. Wollsching-Strobel M, et al. Evolution of web-based training videos provided by the German respiratory league for the correct inhalation technique. Respiration. 2022;101(8):757–65.
95. Park HJ, et al. Video education versus face-to-face education on inhaler technique for patients with well-controlled or partly-controlled asthma: a phase IV, open-label, non-inferiority, multicenter, randomized, controlled trial. PLoS One. 2018;13(8):e0197358.
96. Puah SH, et al. Mobile device: a useful tool to teach inhaler devices to healthcare professionals. BMC Med Educ. 2022;22(1):238.
97. Press VG, et al. Effectiveness of virtual vs in-person inhaler education for hospitalized patients with obstructive lung disease: a randomized clinical trial. JAMA Netw Open. 2020;3(1):e1918205.
98. Harpaz R, et al. Signaling COVID-19 vaccine adverse events. Drug Saf. 2022;45(7):765–80.
99. Husby A, et al. SARS-CoV-2 vaccination and myocarditis or myopericarditis: population based cohort study. BMJ. 2021;375:e068665.
100. Li C, et al. Intravenous injection of coronavirus disease 2019 (COVID-19) mRNA vaccine can induce acute Myopericarditis in mouse model. Clin Infect Dis. 2022;74(11):1933–50.
101. Rzymski P, Perek B, Flisiak R. Thrombotic thrombocytopenia after COVID-19 vaccination: in search of the underlying mechanism. Vaccines (Basel). 2021;9(6):559.
102. McLenon J, Rogers MAM. The fear of needles: a systematic review and meta-analysis. J Adv Nurs. 2019;75(1):30–42.
103. Freeman D, et al. Injection fears and COVID-19 vaccine hesitancy. Psychol Med. 2023;53(4):1185–95.

104. Bouya S, et al. Global prevalence and device related causes of needle stick injuries among health care workers: a systematic review and meta-analysis. Ann Glob Health. 2020;86(1):35.
105. Hosseinipalangi Z, et al. Global, regional and national incidence and causes of needlestick injuries: a systematic review and meta-analysis. East Mediterr Health J. 2022;28(3):233–41.
106. Wang YC, et al. Respiratory care for the critical patients with 2019 novel coronavirus. Respir Med. 2021;186:106516.
107. Viana Martins CP, Xavier CSF, Cobrado L. Disinfection methods against SARS-CoV-2: a systematic review. J Hosp Infect. 2022;119:84–117.
108. Hasija V, et al. Environmental impact of COVID-19 vaccine waste: a perspective on potential role of natural and biodegradable materials. J Environ Chem Eng. 2022;10(4):107894.
109. Friebel C, Steckel H. Single-use disposable dry powder inhalers for pulmonary drug delivery. Expert Opin Drug Deliv. 2010;7(12):1359–72.
110. de Boer AH, Hagedoorn P. The role of disposable inhalers in pulmonary drug delivery. Expert Opin Drug Deliv. 2015;12(1):143–57.
111. Hasija V, et al. The environmental impact of mass coronavirus vaccinations: a point of view on huge COVID-19 vaccine waste across the globe during ongoing vaccine campaigns. Sci Total Environ. 2022;813:151881.
112. Rayhan MRI, Liza JM, Rahman MM. Assessment of COVID-19 vaccination-related medical waste management practices in Bangladesh. PLoS One. 2022;17(8):e0273053.
113. Emami F, et al. Drying technologies for the stability and bioavailability of biopharmaceuticals. Pharmaceutics. 2018;10(3):131.
114. Sharma A, et al. Innovative drying technologies for biopharmaceuticals. Int J Pharm. 2021;609:121115.
115. Kanojia G, et al. Developments in the formulation and delivery of spray dried vaccines. Hum Vaccin Immunother. 2017;13(10):2364–78.
116. Gomez M, Vehring R. Spray drying and particle engineering in dosage form design for global vaccines. J Aerosol Med Pulm Drug Deliv. 2022;35(3):121–38.
117. AboulFotouh K, Cui Z, Williams RO 3rd. Next-generation COVID-19 vaccines should take efficiency of distribution into consideration. AAPS PharmSciTech. 2021;22(3):126.
118. Lee YH, et al. Vaccine-related errors in reconstitution in South Korea: a national physicians' and nurses' survey. Vaccines (Basel). 2021;9(2):117.
119. Samad F, et al. Strategies to reduce errors associated with 2-component vaccines. Pharmaceut Med. 2021;35(1):1–9.
120. Shastay A. Administering just the diluent or one of two vaccine components leaves patients unprotected. Home Healthc Now. 2016;34(4):218–20.
121. Sheng WH, et al. Safety and antibody response of recipients who unexpectedly received undiluted prime dose of BNT162b2 COVID-19 vaccine in Taiwan. J Formos Med Assoc. 2023;122(1):78–81.
122. Hassan AO, et al. A single-dose intranasal Chad vaccine protects upper and lower respiratory tracts against SARS-CoV-2. Cell. 2020;183(1):169–184 e13.
123. Song X, et al. Aerosolized Intratracheal inoculation of recombinant protective antigen (rPA) vaccine provides protection against inhalational Anthrax in B10.D2-Hc(0) mice. Front Immunol. 2022;13:819089.
124. Mettelman RC, Allen EK, Thomas PG. Mucosal immune responses to infection and vaccination in the respiratory tract. Immunity. 2022;55(5):749–80.
125. Duan W, Croft M. Control of regulatory T cells and airway tolerance by lung macrophages and dendritic cells. Ann Am Thorac Soc. 2014;11(Suppl 5):S306–13.
126. Danielsson R, Eriksson H. Aluminium adjuvants in vaccines – a way to modulate the immune response. Semin Cell Dev Biol. 2021;115:3–9.
127. Kool M, Fierens K, Lambrecht BN. Alum adjuvant: some of the tricks of the oldest adjuvant. J Med Microbiol. 2012;61(Pt 7):927–34.
128. Hogenesch H. Mechanism of immunopotentiation and safety of aluminum adjuvants. Front Immunol. 2012;3:406.

129. Xu H, et al. Nasal aluminum (oxy)hydroxide enables adsorbed antigens to induce specific systemic and mucosal immune responses. Hum Vaccin Immunother. 2017;13(11):2688–94.
130. Blaauboer SM, et al. The mucosal adjuvant cyclic di-GMP enhances antigen uptake and selectively activates pinocytosis-efficient cells in vivo. elife. 2015:4:e06670.
131. Facciola A, et al. An overview of vaccine adjuvants: current evidence and future perspectives. Vaccines (Basel). 2022;10(5):819.
132. Bhide Y, et al. Cross-protective potential and protection-relevant immune mechanisms of whole inactivated influenza virus vaccines are determined by adjuvants and route of immunization. Front Immunol. 2019;10:646.
133. Toniolo SP, et al. Excipient selection for thermally stable enveloped and non-enveloped viral vaccine platforms in dry powders. Int J Pharm. 2019;561:66–73.
134. Afkhami S, et al. Spray dried human and chimpanzee adenoviral-vectored vaccines are thermally stable and immunogenic in vivo. Vaccine. 2017;35(22):2916–24.
135. Toniolo SP, et al. Spray dried VSV-vectored vaccine is thermally stable and immunologically active in vivo. Sci Rep. 2020;10(1):13349.
136. Price DN, et al. Design and optimization of a temperature-stable dry powder BCG vaccine. Pharm Res. 2019;37(1):11.
137. LeClair DA, et al. Evaluation of excipients for enhanced thermal stabilization of a human type 5 adenoviral vector through spray drying. Int J Pharm. 2016;506(1–2):289–301.
138. Trenkel M, Scherliess R. Optimising nasal powder drug delivery – characterisation of the effect of excipients on drug absorption. Int J Pharm. 2023;633:122630.
139. Chaturvedi M, Kumar M, Pathak K. A review on mucoadhesive polymer used in nasal drug delivery system. J Adv Pharm Technol Res. 2011;2(4):215–22.
140. Pu Y, et al. A comparison of the deposition patterns of different nasal spray formulations using a nasal cast. Aerosol Sci Technol. 2014;48(9):930–8.
141. Maaz A, Blagbrough IS, De Bank PA. In vitro evaluation of nasal aerosol depositions: an insight for direct nose to brain drug delivery. Pharmaceutics. 2021;13(7)
142. Foo MY, et al. The influence of spray properties on intranasal deposition. J Aerosol Med. 2007;20(4):495–508.
143. Ramvikas M, et al. Chapter fifteen – nasal vaccine delivery. In: Skwarczynski M, Toth I, editors. Micro and nanotechnology in vaccine development. William Andrew Publishing; 2017. p. 279–301.
144. Jiao J, Zhang L. Influence of intranasal drugs on human nasal mucociliary clearance and ciliary beat frequency. Allergy Asthma Immunol Res. 2019;11(3):306–19.
145. Merkus FW, et al. Nasal mucociliary clearance as a factor in nasal drug delivery. Adv Drug Deliv Rev. 1998;29(1–2):13–38.
146. Merkus P, et al. Influence of anatomy and head position on intranasal drug deposition. Eur Arch Otorhinolaryngol. 2006;263(9):827–32.
147. Anderson P. Use of respimat soft mist inhaler in COPD patients. Int J Chron Obstruct Pulmon Dis. 2006;1(3):251–9.
148. Manjaly Thomas ZR, et al. Alternate aerosol and systemic immunisation with a recombinant viral vector for tuberculosis, MVA85A: a phase I randomised controlled trial. PLoS Med. 2019;16(4):e1002790.
149. Low N, et al. A randomized, controlled trial of an aerosolized vaccine against measles. N Engl J Med. 2015;372(16):1519–29.
150. Satti I, et al. Safety and immunogenicity of a candidate tuberculosis vaccine MVA85A delivered by aerosol in BCG-vaccinated healthy adults: a phase 1, double-blind, randomised controlled trial. Lancet Infect Dis. 2014;14(10):939–46.
151. Chan VW, et al. Transmission of severe acute respiratory syndrome coronavirus 1 and severe acute respiratory syndrome coronavirus 2 during aerosol-generating procedures in critical care: a systematic review and meta-analysis of observational studies. Crit Care Med. 2021;49(7):1159–68.

152. Group MA, et al. Safety and immunogenicity of dry powder measles vaccine administered by inhalation: a randomized controlled phase I clinical trial. Vaccine. 2014;32(50):6791–7.
153. Gregoriano C, et al. Use and inhalation technique of inhaled medication in patients with asthma and COPD: data from a randomized controlled trial. Respir Res. 2018;19(1):237.
154. Harb HS, et al. Determinants of incorrect inhaler technique in chronic obstructive pulmonary disease patients. Int J Clin Pract. 2021;75(6):e14073.
155. Kaplan A, Price D. Matching inhaler devices with patients: the role of the primary care physician. Can Respir J. 2018;2018:9473051.
156. Claus S, et al. How can we bring high drug doses to the lung? Eur J Pharm Biopharm. 2014;86(1):1–6.
157. Scherliess R, Etschmann C. DPI formulations for high dose applications – challenges and opportunities. Int J Pharm. 2018;548(1):49–53.
158. Son YJ, Miller DP, Weers JG. Optimizing spray-dried porous particles for high dose delivery with a portable dry powder inhaler. Pharmaceutics. 2021;13(9):1528.
159. Cheng YS. Mechanisms of pharmaceutical aerosol deposition in the respiratory tract. AAPS PharmSciTech. 2014;15(3):630–40.
160. Chow MYT, et al. In vitro-in vivo correlation of cascade impactor data for orally inhaled pharmaceutical aerosols. Adv Drug Deliv Rev. 2021;177:113952.
161. Chaudhary N, Weissman D, Whitehead KA. mRNA vaccines for infectious diseases: principles, delivery and clinical translation. Nat Rev Drug Discov. 2021;20(11):817–38.
162. Chalkias S, et al. A bivalent omicron-containing booster vaccine against Covid-19. N Engl J Med. 2022;387(14):1279–91.
163. Ulmer JB, Valley U, Rappuoli R. Vaccine manufacturing: challenges and solutions. Nat Biotechnol. 2006;24(11):1377–83.
164. Wu S, et al. A single dose of an adenovirus-vectored vaccine provides protection against SARS-CoV-2 challenge. Nat Commun. 2020;11(1):4081.
165. Rosa SS, et al. mRNA vaccines manufacturing: challenges and bottlenecks. Vaccine. 2021;39(16):2190–200.
166. Jackson NAC, et al. The promise of mRNA vaccines: a biotech and industrial perspective. NPJ Vaccines. 2020;5:11.
167. Saluja V, et al. A comparison between spray drying and spray freeze drying to produce an influenza subunit vaccine powder for inhalation. J Control Release. 2010;144(2):127–33.
168. Adali MB, et al. Spray freeze-drying as a solution to continuous manufacturing of pharmaceutical products in bulk. PRO. 2020;8(6):709.
169. Preston KB, Randolph TW. Stability of lyophilized and spray dried vaccine formulations. Adv Drug Deliv Rev. 2021;171:50–61.
170. Lapuente D, et al. Protective mucosal immunity against SARS-CoV-2 after heterologous systemic prime-mucosal boost immunization. Nat Commun. 2021;12(1):6871.
171. Chow MYT, et al. Inhalable neutralizing antibodies – promising approach to combating respiratory viral infections. Trends Pharmacol Sci. 2023;44(2):85–97.
172. van Riel D, de Wit E. Next-generation vaccine platforms for COVID-19. Nat Mater. 2020;19(8):810–2.
173. Tregoning JS, et al. Vaccines for COVID-19. Clin Exp Immunol. 2020;202(2):162–92.

Respiratory Delivery of Probiotics to Improve Lung Health

Alex Seungyeon Byun, Luis Vitetta, Hak-Kim Chan, and Philip Chi Lip Kwok

Abstract Increasing evidence suggests that the lung microbiome is essential for maintaining lung health and homeostasis. Although causality has not been established, differences in its abundance, community richness, and composition are observed in inflammatory lung diseases such as asthma. Furthermore, respiratory infections such as COVID-19 were shown to influence the composition of the lung microbiome. The efficacy of inhaled probiotics to correct this dysbiosis is unknown as the respiratory route of administration is less reported compared to the oral route. The more direct intranasal administration of probiotics may exert a greater protective response against viral respiratory infections compared to the oral route in mice. However, there is a lack of human studies investigating the efficacy of inhaled probiotics. This may be due to the limitations in administration methods in delivering probiotics to humans, as to date, only a nasal spray or a nasal irrigation has been utilised. Nasal sprays require the probiotics to be in a liquid suspension which can affect the longevity of the product. Therefore, investment into developing a more stable probiotic formulation is needed.

Keywords Probiotics · Microbiome · Dysbiosis · Inflammation · Infection

7.1 Microbiome of the Lungs

The common notion that the lungs were sterile and free of bacteria was a discredited self-evident truth. The lungs are constantly exposed to the external environment and the diverse communities of inhaled microbes [1]. Advancements in research with novel culture-independent techniques of microbial identification have provided

A. S. Byun · L. Vitetta · H.-K. Chan · P. C. L. Kwok
Advanced Drug Delivery Group, Sydney Pharmacy School, Faculty of Medicine and Health, The University of Sydney, Camperdown, NSW, Australia
e-mail: philip.kwok@sydney.edu.au

J. K. W. Lam, P. C. L. Kwok (eds.), *Respiratory Delivery of Biologics, Nucleic Acids, and Vaccines*, AAPS Introductions in the Pharmaceutical Sciences 8,
https://doi.org/10.1007/978-3-031-47567-2_7

evidence that the lungs harbour a diverse community of bacteria, fungi, and viruses. Compared to the commensal microbiome of the gastrointestinal tract, the composition of the lung microbiome is subject to a dynamic transient bacterial colonisation and clearance effect due to its physiological and functional anatomy. This is important given that the lungs and intestines share embryological origin and both sites contain mucosa-lined luminal surfaces, with distinctive anatomic and biochemical features. Consequently, the respiratory tract, including the lungs, is constantly exposed to fluctuating microbial transient colonisation and elimination through inhaled air and micro-aspiration, as well as host mucociliary clearance, coughing, and immune defences [2–4]. Subclinical or "silent" micro-aspiration of pharyngeal secretions is common in healthy individuals [5, 6] and changes in micro-aspiration can be observed in inflammatory respiratory diseases [7]. Such changes may contribute to the differences observed in the microbiome of individuals with or without respiratory diseases.

The microbiome is defined as a community of microorganisms that exerts activity within a defined environment [8]. The human lung microbiome is primarily colonised by six phyla, of which Bacteroidetes (Bacteroidota) and Firmicutes (Bacillota) comprise 90% of the biomass. Additional phyla include Proteobacteria (Pseudomonadota), Actinobacteria (Actinomycetota), Fusobacteria (Fusobacteriota) and Cyanobacteria (Cyanophyta). Similar to those of the gut [9], bacterial species from the Bacteriodetes and Firmicutes phyla predominate in the healthy lung [2, 10] and are critical in maintaining immune homeostasis [2, 11, 12] (Table 1). However, despite their similarities, there are differences present at the genus level as, Firmicutes in the gut are comprised of *Lactobacilli* and *Enterococci*, whereas in the lungs, *Veillonella* and *Streptococci* dominate [9, 13]. The lower respiratory tract (LRT) has essentially the same microbiome as that of the upper respiratory tract (URT) because they are anatomically contiguous [3, 14]. To compare the bacterial communities present in the upper and lower airways, Charlson et al. [14] obtained oro-/nasopharyngeal swabs and bronchoscopy samples from the glottis and serial bronchoalveolar lavage (BAL) from six healthy individuals, respectively. This revealed that the microbiome of the lungs is an extension of that of the URT but with a lower biomass [14]. This may again be due to the anatomy of the respiratory tract being a continuous mucosal surface. As the

Table 1 Gut and lung microbiome composition in the order of abundance

	Gut [10]		Lung [13]	
Highest relative abundance	Firmicutes	77.8%	Bacteroidetes	43%
	Bacteroidetes	12.5%	Firmicutes	24.5%
	Proteobacteria	4.9%	Proteobacteria	20.3%
	Actinobacteria	4.1%	Actinobacteria	6.6%
	Other phyla e.g., Verrucomicrobia, Euryarchaeota, Fusobacteria	<0.7%	Fusobacteria	4.1%
Lowest relative abundance			Other phyla including Cyanobacteria	<1.5%

oropharynx produces approximately 2 L of saliva per day, consequently the microbiome of the URT resembles that of the lung through micro-aspiration [4]. The characterisation difficulties encountered are due to obtaining sufficient uncontaminated samples that are strictly from the LRT. Sequencing the obtained samples is challenging due to the low biomass at this site [2, 14]. Access to the lungs requires bronchoscopes to traverse through the oral or nasal route, which has the potential risk of bacterial sample contamination from various parts of the airway that are not intended to be sampled. Several studies have implemented technical control to prevent cross-contamination of oral or nasal samples by the BAL microbiome [3, 7]. Notwithstanding, potential cross-contamination should be noted when inspecting lung microbiome samples.

7.1.1 Gut-Lung Axis

A crosstalk between the microbiome of the gastrointestinal tract and the respiratory tract exists and is termed the gut-lung axis, which has been extensively studied and reviewed [11, 12, 15]. Shifts in the commensal microbiome of the gut have been reported to reflect immune responses in the lungs, with studies that administered oral probiotic formulations reporting positive effects on the respiratory immune system during an infection [16–19]. Host microbiome interactions that progress to intestinal dysbiosis can manifest themselves as a dysbiotic pro-inflammatory insult at a distant site (e.g., the lungs). For example, chronic obstructive pulmonary disease (COPD) frequently occurs with inflammatory bowel diseases [11, 15]. Although the underlying mechanisms have not been clarified, epidemiological studies indicate that the incidence of COPD is strongly associated with that of Crohn's disease [15]. This further supports the hypothesis that there is a link between the intestinal microbiome and those in the respiratory tract, both in immunological and pathogenic dispositions.

7.1.2 Asthma and Differences in the Lung Microbiome

Asthma is a chronic inflammatory airway disease that affects an estimated 262 million people worldwide as of 2019 [20]. The clinical perception that there is a predisposition to developing asthma, remains unclear, despite various pathogenic hypotheses proposed over the years. The hygiene hypothesis was first proposed in 1989 [21] when Strachan suggested from reported epidemiological studies that a lower incidence of childhood infections was associated with higher standards of personal sanitation. Over several decades, early exposure to microbes from lifestyles adapted to rural-based dwellings [22], contact with older siblings [23, 24], and modern day-care attendance [24] have significantly influenced the diversity and maturity of the gut microbiome that has led to significantly decreased risks of asthma [11, 12]. Furthermore, antibiotic use in the first year of life seems to have a significant adverse effect on the gut microbiome by increasing the risk of developing asthma within the first 3 years of life [25]. Hence, early life infectivity for the neonate is important for developing immunological tolerance and metabolic homeostasis, which significantly reduces the risk of atopic disease development [26].

Research shows that the composition of the lung microbiome differs between an asthmatic patient and a non-asthmatic patient. However, whether the difference contributes to the pathogenesis of asthma is unclear. The causality between asthma and the observed changes in the lung microbiome has not been clearly established. There are gaps in current knowledge in establishing the role of the lung microbiome and its compositional changes in patients with asthma or other respiratory conditions [27]. The landmark paper by Hilty et al. [28] overthrew the long-held dogma that the lungs are sterile. Despite being a small study, they have laid the foundation that the composition of the microbiome differed between non-asthmatic and asthmatic subjects. There was a strong association between the prevalence of microbial species from the Proteobacteria phylum and asthma patients [29], whereas bacterial species from the Firmicutes phylum were more common in non-asthmatic individuals, confirming similar impressions from other studies [30–32]. Huang et al. [30] further demonstrated that lower Asthma Control Questionnaire scores corresponded with the relative abundance of Proteobacteria. This finding supports the earlier report that bronchial hyperresponsiveness in asthma correlates with the composition and diversity of the lung microbiome [31]. It should be noted that the subjects recruited differed between the studies as not all individuals were managed with inhaled corticosteroids (ICS). Nonetheless, the results showed consistency in increased Proteobacteria in asthmatic patients, hence it is unlikely that ICS use in mild-to-moderate asthma affects the microbiome. In medication-controlled severe asthmatics, an increase in Actinobacteria, especially bacteria from the *Klebsiella* genus, correlated with improved asthma management and bronchial epithelial gene expression of the FK506 binding protein, a marker of ICS responsiveness [30]. Furthermore, investigation into the microbiome of ICS-resistant and ICS-sensitive patients revealed that there was an increase in Proteobacteria for both groups but an increase in Actinobacteria was only observed in ICS-resistant patients compared to non-asthmatic patients [33]. In comparison between ICS-resistant and ICS-sensitive patients, the differences in the lung microbiome were at the genera level, with expansions in *Neisseria*, *Haemophilus*, *Simonsiella*, *Campylobacter*, *Leptotrichia*, *Tropheryma*, *Leuconostoc*, and *Megasphaera* observed in ICS-resistant patients [33]. *Haemophilus parainfluenzae*, an organism that was unique to the airways of ICS-resistant patients, induced a dose and time-dependent activation of the p38 mitogen-activated protein kinase (MAPK) pathway that consequently reduced the corticosteroid responses of BAL macrophages [33]. However, such effects were not seen with the commensal airway organism *Provotella melaninogenica*. Therefore, differences in airway microbiome composition may be indicative of the responsiveness to ICS treatment.

Compared to the saline-treated group, dexamethasone (DEX) treated C57BL/6 mice displayed phylogenetic shifts in gut microbiome with expansions in Firmicutes and especially Actinobacteria in 10 days of treatment [34]. Additionally, mucin gene expression Muc2 was significantly decreased in DEX-treated mice [34]. Consequent faecal transplantation of DEX-treated mice to germ-free mice led to increased expression in Muc1 and Muc2 genes, suggesting that gut microbiome may regulate mucin production [34]. However, the mechanisms of how corticosteroids (CS) affect the composition of the gut microbiome are unknown [34, 35]. Due to the gut-lung axis, it is possible that the changes in the gut microbiome as a result of CS administration could also be reflected in the lungs and vice versa as a recent systematic review concluded that CS could influence the composition of the lung microbiome [36]. However, it should also be noted that the current literature is limited and confounding and therefore further research is necessary for a deeper understanding of the effects of both inhaled and systemic CS on the respiratory microbiome [36].

7.1.3 Different Sampling Methods of the Lung Microbiome

Discrepancies exist in the findings from severe asthmatics with reports that show an increase in the abundance of bacterial species from the Firmicutes phylum with a decrease in the abundance of bacterial species from the Bacteriodetes and Fusobacteria phyla [32]. A possible explanation for these discrepancies may be attributed to the different sampling methods. Protected bronchial brushing utilised by Huang et al. [30] was designed to limit contamination when passing through the tracheobronchial tree [37]. Sputum sampling is commonly used for determining the aetiology of LRT infections. Naturally, as sputum involves spitting out from the mouth, it would be more prone to contamination by the oral microbiome. Zhang et al. [32] specified the care taken to avoid this contamination by implementing oral washes and nose blowing. However, there may be patient variation in the quality of the sputum obtained. It was previously reported that only 54% of the collected sputum samples were of good quality [38]. The comparison of the commonly used sampling methods showed that the proportion of Proteobacteria increased and Firmicutes diminished in the order of oral wash, small-volume BAL, protected BAL and protected specimen brush sampling [39]. The human oral microbiome confirms that Firmicutes is the predominant phylum (36.7%) in the mouth [40]. Zhang et al. [32] also confirmed that the proportion of Proteobacteria increased in non-severe asthma and that non-asthmatic subjects had Bacteriodetes occupying the largest proportion, suggesting that sputum sampling is of similar quality to that of brush sampling. However, it is difficult to discern due to the broad classification of microbiome into phyla, leading to ambiguity of data. It is possible that the differences lie in the location of sampling within the lung. Potential spatial variation of the lung microbiome was studied by comparing samples from the lingula of the lung, right middle lobe, right and left upper lobe and supraglottic space using BAL or a protected-specimen brush (PSB) [41]. In healthy subjects, the variation between BAL and PSB was insignificant, and intrapulmonary variation was less notable

[14, 41] than interpersonal variation. This may be due to the individual's lung microbiome almost being an extension of their URT community [14, 41]. Then, as the lungs are constantly exposed to the environment and display dynamic bacterial communities, could geography possibly play a part in determining the individual's microbiome structural profile?

7.1.4 Geographical Differences

Currently, whether geographical differences influence the structure of the lung microbiome is yet to be elucidated. Morris et al. [42] recruited 64 patients across eight cities in the USA for the purpose of determining the difference between non-smokers' and smokers' lung microbiomes but they did not specifically study the potential differences between subjects from different geographical areas. Significant differences in bacterial diversity and community structure of the gut were observed, depending on the geographical region and study participant age [43]. Gut diversity is highly dependent on dietary practices, which makes it highly likely that intestinal microbiome diversity differences could be dependent on geographical location, as demonstrated by the significant variability observed between USA and non-USA gut microbiome studies [43]. The likelihood that differences in the diversity and structure of the lung microbiome according to geographical location cannot be dismissed because the airway lumen is continuously exposed to the external environment. Therefore, given that regardless of the disease state the lung microbiome is in a dynamic state, it is expected that there would be observed microbiome discrepancies in the lungs of patients with respiratory conditions.

7.1.5 Dysbiosis of the Lung Microbiome and Respiratory Diseases

Microbial dysbiosis in the lungs is present in asthma, and although not covered in this review, it is also present in other inflammatory diseases such as COPD [13, 44, 45] and cystic fibrosis [44, 46, 47]. More recently, dysbiosis was also observed in respiratory infections, such as influenza in the gut [48, 49] and the coronavirus disease (COVID-19) [50] in the lungs. Sublethal influenza A virus infection resulted in significant but a temporary reduction in the bacterial content in the gut of the mice, a change that was not observed in the URT or LRT [48]. In patients with differing severity of COVID-19, the relative abundance in Firmicutes and Actinobacteria increased whereas Bacteroidetes and Proteobacteria decreased in the URT compared to healthy individuals [50]. In the LRT, patients with fatal COVID-19 showed a higher abundance of Bacteroidetes and Proteobacteria compared to those with severe disease [50]. At the genus level, there was an increase in *Streptococcus* as the disease severity increased and increased levels of *Veillonella*, *Staphylococcus* and *Actinomyces* compared to patients without COVID-19 [50]. These genera contain opportunistic pathogens such as *S. aureus* which was the predominant pathogen responsible for co- and secondary infections in patients admitted to hospital with COVID-19 [51]. Although it may seem worthwhile to correct the dysbiosis with probiotics, a systematic review of oral probiotics found that current literature up to 2014 poorly supported this view [52]. Many analysed studies were ambiguous in stating which strain was tested as they only reported the genus or species of the probiotics [52]. The need to single out which strain is beneficial was highlighted [52]. To date, inhaled probiotics have not been administered to correct dysbiosis in the lung microbiome and hence further research is needed.

7.2 Effects of Probiotics on Respiratory Infections

Probiotics are defined as quantifiable live bacteria with evidence for health benefits at either a strain or group level, according to the International Scientific Association for Probiotics and Prebiotics [53]. However, this review will also include non-viable counterparts of probiotic strains in the definition. Intranasal administration of probiotics has resulted in positive outcomes in respiratory infections with animal models [16, 54–64]. Over the past decade, *Lactobacillus* has been shown to be effective in improving survival of mice infected with the influenza virus (IFV) [16, 54, 56–59, 64, 65]. Harata et al. [54] first demonstrated that non-viable *L. rhamnosus* GG was effective in improving clinical outcomes. Furthermore, heat-killed strains including *L. casei* DK128 [57], *L. pentosus* S-PT84 [58] and *L. plantarum* NCIMB 8826 [60] demonstrated effectiveness against subtype H3N2 and H1N1 of IFV and also against pneumonia virus of mice, which is closely related to RSV. Tomosada et al. [55] confirmed that non-viable probiotics were equally effective as viable cells in protecting the host against respiratory syncytial virus (RSV), a characteristic that was *Lactobacillus* strain specific.

When challenged with poly(I:C) in vitro, heat-killed *L. plantarum* MPL16 and CRL1506 resulted in no changes in IFN-β, IL-6 and chemokines compared to non-treated Calu-3 cells [66]. Also, in the case of severe acute respiratory syndrome coronavirus (SARS-CoV-2), unlike their viable counterparts, these heat-killed strains failed to reduce the viral titre nor reduced lactate dehydrogenase (LDH) levels, a biomarker for tissue damage. This observation indicates that non-viable *L. plantarum* MPL16 and CRL1506 do not exert protection against SARS-CoV-2 and the ability to immunomodulate and protect against respiratory diseases is strain-specific [66]. Furthermore, heat-killed *L. plantarum* MRL18 and CRL1506 were unable to exert an immunomodulatory response post poly(I:C) challenge unlike their live counterparts [66]. Similar effects were also observed when tested against IFV [56, 64, 65]. Although heat-killed *L. rhamnosus* CRL1505 increased the resistance of infant mice to RSV as effectively as viable probiotics, the greatest effect in reducing lung injury was observed with the live strains [55]. Based on this evidence, an experimental posit should be employed to determine whether viable cells increase adverse effects and whether this trade-off is worthwhile.

Further evidence suggests that intact probiotic bacterial cells may not be necessary to induce a positive immunological effect to clear respiratory infections. Cell wall fractions of *Bifidobacterium longum* 35624® have been shown to reduce the viral load in lung tissue compared to controls, additionally improving survival in a lethal influenza infection mouse model through reduced acute lung injury [61]. The peptidoglycan part of the cell wall of *L. rhamnosus* strain CRL1505 exerted specific positive immunomodulatory effects with increased resistance against respiratory pathogens such as *Streptococcus pneumoniae* [67, 68] whilst limiting lung damage [67]. The potential of immunomodulatory activity of unviable whole cells and/or their fragments suggests the possibility of their use in immunocompromised individuals to avoid inadvertent infection from viable probiotic bacteria.

7.2.1 Possible Mechanisms of Probiotics Against Viral Infections

There are no established mechanisms that explain how probiotic strains such as *Lactobacillus* and *Bifidobacterium* benefit the host in viral infections. It is possible that mechanisms are strain-specific. In an IFV infection murine model, *L. rhamnosus* GG stimulated the activity of natural killer (NK) cells which are responsible for quickly subduing pathogens [54]. The authors have suggested that the concurrent increase in IL-1β, TNF and MCP-1 mRNA in the lungs of *L. rhamnosus* GG-treated mice strengthens the argument that *L. rhamnosus* GG utilises NK cells to protect against IFV. This is because TNF and MCP-1 activate NK cells and IL-1β with IL-12 co-stimulates IFN-γ which then encourages NK cell proliferation [69].

Youn et al. [16] and Tonetti et al. [64] reported that the administration of *L. rhamnosus* (unspecified and CRL1505 respectively) stimulated the production of mucosal secretory IgA in the lungs. IgA exerts protective effects against IFV [70] by binding to viruses directly to prevent entry into the mucosal epithelium thereby reducing the viral titre in the lungs [64]. In IFV-infected mice, significant increases in IL-6 and TNF-α positively correlated with lung inflammation [16, 18]. The addition of *L. rhamnosus* led to consequent reductions of IL-6 and TNF-α, thereby reducing inflammation caused by IFV [16, 64]. IL-6 and TNF-α are typical immune responses that are measured for acute infections when examining toll-like receptor 2 (TLR2) [71]. Percopo et al. [60] have also suggested that *L. plantarum* NCIMB 8826 interacts with TLR2 and nucleotide-binding oligomerisation domain-containing protein 2 (NOD2) but does not induce protection through stimulation of IL-10 and does not need to signal type 1 IFN downstream. They have concluded that further consideration of cross-signalling between TLR2 and NOD2 is required. Kolling et al. [68] also suggested that viable cells may induce uncontrolled immune responses as peptidoglycan cell walls in Gram-positive bacteria activate TLR2 coupling with TLR6 to stimulate the production of proinflammatory cytokines such as IL-8 and TNF-α. Such a result could be dangerous for immunocompromised patients [68]. Heat-inactivation of *L. rhamnosus* CRL1505 may inadvertently change molecular structures on the surface of the cell, therefore making it less efficiently recognised by pattern recognition receptors [68]. As its exact mechanisms are unknown, a reduced immune response may be a trade-off with efficacy for non-viable cells.

7.2.2 Safety of Probiotics in the Human Respiratory Tract

Animal studies have shown promising results that justify the use of probiotic bacteria in the management of respiratory infections [16, 54–64]. Currently, clinical trials on the intranasal delivery of probiotics are limited. However, oral delivery of probiotics in humans has been shown to influence the local immune system in the lungs and hence positively affect respiratory infections [72, 73]. It was not clear from animal studies whether adverse effects were present and hence the safety of intranasal delivery of probiotics was investigated [74]. The prospective study by Endam et al. [74] reported no serious adverse events when *L. lactis* W136 in 0.9% NaCl was delivered by a nasal irrigation apparatus for chronic rhinosinusitis. The minor adverse reactions included headache, migraine, nasal congestion, cold sore, throat pain, and gastroenteritis. Some of these events were reported to be pre-existing conditions. Since no placebo control was used in this study, it was difficult to conclude whether the adverse reactions were due wholly to the probiotic bacteria or from the conditions themselves given the ineffective treatment. However, as administration of the probiotic resulted in reported progressive symptom improvement, the probiotics were unlikely to be responsible for those side effects. As concluded by the safety document released by the USA government [75], adverse events were vague in nature or unreported in many published studies. A more recent randomised controlled trial of a nasal spray containing *Bacillus subtilis* ANA4 and *Bacillus clausii* ANA39 in 0.9% NaCl (LiveSpo® Navax) stated that both the placebo and treatment were well-tolerated in children without signs of nasal mucosa irritation, local bacterial infection, vomiting, or diarrhoea [76]. In another recent nasal spray study containing *L. casei* AMBR2, safety and tolerability were thoroughly evaluated by recording symptoms and biomarkers, as well as conducting physical examinations of the URT by an ear, nose, and throat specialist [77]. No physical signs of acute inflammation were present, which was supported by normal levels of C-reactive protein in 15 out of 20 volunteers. Nasal symptoms were only periodic and did not last the whole duration of the study [77]. Along with previous studies [77, 78] and the fact that *Lactobacillus* spp. have been listed under the qualified presumption of safety by the European Union (EU) [79], De Boeck et al. concluded that *L. casei* AMBR2, *L. rhamnosus* GG and *L. plantarum* WCFS1 were safe as a throat spray and did not monitor for adverse reactions [80]. Three individuals discontinued the treatment in the intervention group due to the taste of the spray, which was a factor that affected patient compliance. No safety data on probiotics delivered into the airways is yet available. Their effects on the microbiome of the respiratory tract are also unknown. There is a need to further study the safety profile of probiotics in the respiratory tract if the aim is to realise the potential of probiotics for respiratory infections.

7.3 Intranasal Delivery of Probiotics

Despite the ambiguity of the safety data for probiotics, the effectiveness of probiotics is well-demonstrated [16, 54–64] (Table 2). Although the data is replete for orally administered probiotics for indirect respiratory conditions, a direct mode of delivery to the lungs could enhance therapeutic efficiency. The existence of a gut-lung axis indicates the immunomodulating capacity of the gut microbiome that extends to the lungs [11], and hence oral probiotics can protect against respiratory conditions and improve survival in mice [16, 59], as well as improve clinical outcomes in humans [19, 80]. However, intranasal instillation was superior to oral delivery in mice [16, 59]. In mice challenged with the influenza virus (A/NWS/33 H1N1), intranasally delivered *L. rhamnosus* achieved a higher survival rate than oral administration [16]. Furthermore, the same effect was demonstrated with *L. plantarum* DK119, for which the efficacy via the oral route was less pronounced than intranasal delivery [59]. Intranasal *L. paracasei* NCC2461 had higher efficiency in reducing inflammatory chemo- and cytokine eotaxin and IL-5 in allergic lung inflammation than the intragastric route [17]. The commonly used dose for intranasal delivery in animal studies was 10^8–10^9 colony-forming units (CFU) daily [16, 17, 55–57, 59–61], whereas at least 10^9–10^{10} CFU daily were administered orally [72, 73, 81–83]. In general, oral delivery requires higher doses as drugs are often prone to first-pass metabolism. However, probiotics are not absorbed into the systemic circulation and hence are not subjected to hepatic metabolism. Instead, orally delivered probiotics need to survive the transit through the gastrointestinal tract, especially in gastric acid. It is thus less efficient than direct intranasal or pulmonary delivery for respiratory conditions. This efficiency was demonstrated by Fangous et al. [84] with a cocktail of *Lactobacillus* species delivered to mice by intratracheal administration. A low dose of 9×10^6 CFU was enough to significantly reduce the bacterial load of *P. aeruginosa* in the murine lungs [84]. Respiratory delivery is non-invasive, efficacious at lower doses, and may elicit a faster onset of therapeutic action.

Currently, data on airway delivery of probiotics in humans is limited, possibly due to the complexities in formulating orally inhaled probiotics compared to nasal instillation in animal studies. Recently, Tran et al. [76] investigated the effectiveness of probiotics (*Bacillus subtilis* ANA4, *Bacillus clausii* ANA39) nasal sprays in improving clinical respiratory symptoms of RSV in children. The outcomes were promising, with symptoms having a significantly shorter period in the treatment group with improved efficacy in RSV load clearance by 53-fold compared to the

Table 2 Intranasal delivery of probiotics in mice

Strain	Delivery/ mice	Dose	Infection	Outcome	Reference
Lactococcus lactis NZ9000	Intranasal Swiss albino mice	10^8 cells daily for 2, 5, or 7 days	*S. pneumoniae* lung infection	5-day administration increased clearance rate of *S. penumoniae* from lungs Reduced lung injury in probiotic-treated mice	[62]
L. rhamnosus GG (heat-killed)	Intranasal BALB/c mice	200 μg of lyophilised *L. rhamnosus* GG daily for 3 days (the number of cells or CFU in the administered dose was not specified)	IFV A/PR8/34 (H1N1)	Improved survival of treated mice by 40% Pulmonary IL-1β, TNF mRNA expression is significantly higher in the probiotic group	[54]
L. pentosus S-PT84 (heat-killed) (food-derived)	Intranasal BALB/c mice	200 μg of lyophilised *L. pentosus* S-PT84 daily for 3 days (the number of cells or CFU in the administered dose was not specified)	IFV A/PR8/34 (H1N1)	Intranasal priming for 4 days resulted in a significantly higher survival rate (94.1%) on day 14 compared to the control (58.8%) Treated mice had reduced viral load on day 7 compared to untreated mice	[58]

(continued)

Table 2 (continued)

Strain	Delivery/ mice	Dose	Infection	Outcome	Reference
L. rhamnosus *L. fermentum*-1 (live and heat-killed)	Intranasal Oral BALB/c mice	10^8 CFU daily for 21 days	IFV A/NWS/33 H1N1	Intranasal route had a higher survival rate than the oral route (*L. rhamnosus*) Live cells resulted in a higher survival rate compared to heat-killed cells in both intranasal and oral routes (*L. rhamnosus*) Level of protection is strain-specific e.g. *L. fermentum*-1 showed a 100% survival rate when administered intranasally	[16]
L. rhamnosus CRL1505, *L. rhamnosus* CRL 1506 (live and heat-killed)	Intranasal BALB/c mice	10^8 CFU daily for 2 days	RSV	Live and heat-killed CRL1505 strain was equally effective in reducing viral load by at least 2 log PFU/g Heat-killed CRL1506 was not as effective in reducing the viral load as live cells Reduced lung injury observed for all tested strains but live CRL1505 demonstrated the greatest effect	[55]

(continued)

Table 2 (continued)

Strain	Delivery/ mice	Dose	Infection	Outcome	Reference
L. plantarum DK119 (food-derived)	Intranasal Oral BALB/c mice	10^9, 10^8, or 10^7 CFU daily for 1 day	IFV A/PR8/34 (H1N1), IFV A/P82 (H3N2)	Intranasal administration of DK119 improved clinical outcomes compared to untreated mice regardless of dose Anti-viral effect is dose-dependent Oral administration is less effective (by 400-fold) in reducing viral load compared to intranasal (10^9 CFU)	[18]
L. rhamosus CRL1505 (live and heat-killed)	Intranasal BALB/c mice	10^8 CFU daily for 2 days	IFV A/PR8/34 (H1N1)	Improved survival rate of treated mice by 20% (heat-killed) and 30% (live) compared to untreated mice Both live and heat-killed CRL1505 reduced lung damage but live cells had greater effect Both live and heat-killed strains significantly reduced viral load	[56]

(continued)

Table 2 (continued)

Strain	Delivery/ mice	Dose	Infection	Outcome	Reference
L. plantarum NCIMB 8826 (ATCC BAA-793) (live and heat-killed) *L. reuteri* F275 (ATCC 23272) (live and heat-killed)	Intranasal BALB/c, C57BL/6 mice	10^9, 10^8, or 10^7 CFU daily on day −14 and day −7 prior to infection	Pneumonia virus of mice	100% survival of mice that received 10^8 CFU on day 1 (*L. plantarum*) Live and heat-killed *L. plantarum* have a similar levels of effect 100% survival of mice that received 10^9 CFU of either live or heat-killed *L. reuteri* on day 1 Similar observations made in C57BL/6 mice	[60]
L. casei DK128 (heat-killed) (food-derived)	Intranasal BALB/c mice	10^9, 10^8, or 10^7 CFU daily on day −4 and day −1 prior to infection	IFV (H3N2)	10^9 CFU/daily boasted 100% survival and no weight loss in mice Dose-dependent effect 10^9 CFU/day resulted in an 18-fold reduction in viral load compared to untreated at day 7 Reduction in BAL IL-6, TNF-a	[57]
B. longum 35624®	Intranasal BALB/c mice	Cell: 10^9 CFU Cell wall fraction: 150 mg/mL on alternate days for 5 days	IFV A/PR8/34 (H1N1)	Both intact cell and cell wall fraction of strain 35624 reduce viral load at day 5 of infection Improved survival observed with a cell wall fraction of 35624 Reduction of acute lung injury in cell wall fraction of 35624	[61]

(continued)

Table 2 (continued)

Strain	Delivery/ mice	Dose	Infection	Outcome	Reference
L. rhamnosus CRL1505 (live and heat-killed)	Intranasal BALB/c mice	10^9 cells fortnightly for 4 weeks	IFV A/PR8/34 (H1N1)	Both live and heat-killed CRL1505 are equally effective in improving levels of all anti-IFV antibodies studied Both types of CRL1505 significantly increase IFN-g, IL-4 and IL-10 Treated mice showed higher levels of anti-IFV IgA and IgG antibodies compared to untreated mice	[64]
Bacillus subtilis 3, *B. subtilis* 281, *B. amyloliquefaciens* 298 (food-derived)	Intranasal ICR mice	10^7 CFU daily for 1 day	Lung inflammation caused by *S. pneumoniae*, *B cereus*, *E. coli*, *A. alternata*, *C. sphaerospermum P. aeruginosa*, *S. enterica subspecies*, *Enterica serovar Typhimurium*, *L. innocua*	All strains inhibit the growth of *E. coli* (ATCC 25922), *B. cereus* (ATCC 11778), *A. alternata*, *C. sphaerospermum* *B. subtilis* 281 and *B. amyloliquefaciens* 298 inhibit *E. coli* (ATCC 8739), PA, *S. enterica* subsp. *Enterica serovar Typhimurium*, *B. cereus* (ATCC 27348) *B. subtilis* 3 inhibit *L. innocua* IL-10 significantly increased in RAW264.7 macrophage cell line in all strains	[63]

0.9% NaCl control group [76]. Similarly, individuals treated with nasal irrigation with *L. lactis* W136 for their chronic rhinosinusitis displayed progressive symptomatic improvement over 14 days [74]. However, for allergic rhinitis, high-dose intranasal administration of *L. rhamnosus* SP1, *L. paracasei* 101/37, and *L. lactis* L1A resulted in no significant difference in terms of quality of life, nasal symptom score, peak nasal inspiratory flow, and did not change the immune response [85]. It is likely that the choice of probiotic strain/s was the major determinant of the effects as the extent of immunomodulation is dependent on the probiotic strains employed [17, 55, 60, 65, 66, 86]. None of the three strains used in the nasal spray were previously studied for allergic rhinitis, let alone for respiratory conditions [87–89]. Studies on respiratory delivery of probiotics in humans are limited and the translatability of animal data to humans requires investigation.

7.4 Formulating Probiotics for Lung Delivery

The different modes of delivery of active pharmaceutical ingredients to the respiratory tract include nasal sprays, pressurised metered-dose inhalers, dry powder inhalers, nebulisers, and soft mist inhalers. There have been preliminary studies in formulating nasal sprays to target respiratory tract conditions such as viral infections [76] and rhinosinusitis [74], but nasal sprays traditionally are most appropriate in targeting local nasal conditions. To date, there are no reported studies delivering probiotics directly into the lungs.

The advantages of dry powder formulations are that they do not require refrigeration, which facilitates transport and storage. Methods of powder production include freeze drying and spray drying, used extensively in the food industry and applicable to probiotics. Although freeze drying is more preferable than spray drying for probiotics, its productivity is lower and consumes twice as much energy [90]. To date, spray drying has been solely utilised for improving the storage of probiotics [90, 91]. Using spray drying to produce inhalable probiotic powders and testing their aerosol performance have not yet been performed. The main concern of using spray drying for probiotics includes probiotic cell stress caused by heat and dehydration, which can greatly reduce viability [90–95]. Drying temperature is a key parameter that determines the viability. The survival of probiotics has been demonstrated to be affected by the outlet temperature [96]. Generally, the lower the outlet temperature, the greater the viability [96–98]. Outlet temperatures below 70 °C can yield more than 95% of viable bacterial cells [99]. Furthermore, viability during spray drying can be further improved by pre-treating the culture and by adding protectant excipients to encapsulate the bacteria. Mild heat treatment under 60 °C prior to spray drying was shown to improve the viability [96, 100]. However, such treatments can be strain-specific as no improvement was observed for *L. acidophilus* A9, *L. paracasei* A13, and *L. plantarum* com [100], whilst heat pre-treatment effectively reduced cell death for *L. casei* Nad and *L. plantarum* 8329 [100], and *L. salivarius* NRRL B-30514 [96]. It should be noted that the increased tolerability induced by pre-treatment with heat is strain-dependent and its exact mechanisms are unknown.

The addition of inorganic salts can enhance the thermotolerance of spray-dried probiotics. First, calcium ions are essential in cell physiology and are responsible for processes such as signalling, cell division [101], development, and fitness [102]. Huang and Chen [103] demonstrated that 5 and 10 mM of calcium ions increased the heat resistance of *L. casei* Zhang and *S. thermophilus* ND03, thereby improving the viability of these species. Similarly, 1 mM of calcium ions spray-dried with *L. rhamnosus* GG and trehalose or lactose increased the viability by six-fold [93]. Also, given that the addition of calcium salts did not alter powder morphology, crystallinity and infrared transmittance spectrum [93], they may be useful excipients for spray drying probiotics.

Another method to minimise cell death during spray drying includes the addition of protectants such as disaccharides and divalent ions. It is well known that sugars can stabilise the cellular membrane and hence protect cells from dehydration. The membrane phase transition temperature (T_m) can be decreased by stabilisation of the phospholipid cell membrane by the hydrogen-bond interaction with the sugars [92]. Hence having a glass transition temperature (T_g) higher than T_m would effectively achieve better protection [90, 92, 96]. Popular disaccharides for this purpose are lactose [96, 99, 104] and trehalose [93, 96, 99, 105], as the T_g of lactose and trehalose is 101 °C [106] and 107 °C [107], respectively. Although lactose is an approved inhalation excipient, trehalose is yet to be approved by the USA Food and Drug Administration (FDA), despite its Generally Recognised as Safe (GRAS) status and wide use in non-inhaled pharmaceutical and food products [108]. On the other hand, sucrose has a much lower T_g at 60 °C [107] and is less effective in protecting probiotics during spray drying. This characteristic has been demonstrated by Zhang et al. [96] in that lactose-containing probiotic mixtures displayed a higher viability compared to sucrose-containing ones after spray drying. Similarly, spray-dried lactose- or trehalose-containing formulations had significantly higher probiotic viability than those with sucrose [105]. Therefore, current evidence supports the utilisation of common disaccharides as protectants.

Although particle size has not been reported separately in the current publications, scanning electron microscopic images showed that the spray-dried particles were generally 10 μm in diameter or smaller [93, 105]. It should be noted that *Lactobacillus*, which is a common probiotic strain, is generally 1–9 μm in length, depending on the species [109]. This size is at least 10 times larger than smaller biological organisms, such as bacteriophages, which were spray-dried into suitable powders for inhalation (Fig. 1). This factor could be a challenge as the optimal

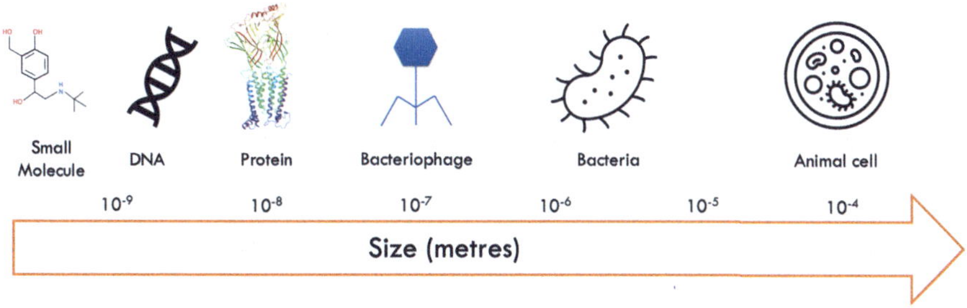

Fig. 1 Comparison of small molecules, nucleic acids, and biologics

particle size for pulmonary deposition is less than 5 μm, further modifications should be made to produce inhalable powders. Another important factor to consider is whether spray drying alters the functionality of the probiotic bacterial strain. Iaconelli et al. [95] reported that spray drying decreased the esterase activity of *L. plantarum* CNRZ 1997, *B bifidum* CIP 56.7, and *L. zeae* CNRZ 2268 and that the extent of this reduction was strain-specific. Furthermore, although the spray-dried bacterial cells did not affect the production of IL-10 (an anti-inflammatory cytokine), they decreased the production IL-12 (a pro-inflammatory cytokine) [95]. However, this effect was not unique to spray drying. It was also observed in other drying processes including freeze drying and air drying [95]. Similarly, Jokicevic et al. [105] reported that spray drying decreased the antimicrobial effects of *L. casei* AMBR2 against *S. aureus* but not against *H. influenzae*. The main mechanism of action against respiratory infection of probiotics was attributed to immunomodulation. The effect of spray and freeze drying on changing the capabilities of probiotic strains needs further investigation.

7.5 Conclusion

The microbiome of the lungs is an extension of that in the URT and is transient with a low concentration. As the lungs are exposed to the environment, interpersonal variations are evident. Upon comparing individuals with and without lung disease, those with inflammatory lung diseases (e.g., asthma) express a different lung microbiome, one that is high in microbial density and low in microbial diversity. However, it is unclear whether the variation in the commensal microbiome led to the disease or whether the disease altered the commensal composition. Nevertheless, strain-specific immunomodulatory capabilities of probiotics have opened novel therapeutic options for the administration of probiotics to be used for managing respiratory infections. The traditional oral route was shown to be effective against lung infections. However, direct delivery to the nose and lungs has been shown to be a more plausible potent mode of therapeutic delivery. The pulmonary route has been extensively studied in animal models but less so in human trials. Currently, safety data on probiotics are not well defined, albeit the FDA has labelled probiotic bacteria with a GRAS grading for the *Lactobacillus* and *Bifidobacterium* genera strains currently in use. Consequently, investigators have dismissed further examining the safety on the presumption that they are also deemed safe by the EU. Future studies need to examine the safety profile of the probiotic strains to strengthen their position as a drug product. Finally, spray drying of probiotics has been commonly utilised in the food industry mainly for storage. The high heat and drying stresses from spray drying greatly decrease probiotic survival. To improve the viability, pre-treatment with mild heat was shown to increase the survival of bacterial cells but this effect is strain-specific. The addition of excipients such as divalent cations (e.g., calcium ions) and disaccharides (e.g., lactose and trehalose) improved the viability after spray drying. The foundations for formulating inhalable probiotic powders have been set so their development is envisaged in the near future.

Acknowledgement The authors received no financial support for the research, authorship, and publication of this article.

References

1. Dickson RP, Erb-Downward JR, Martinez FJ, Huffnagle GB. The microbiome and the respiratory tract. Annu Rev Physiol. 2016;78:481–504. PubMed PMID: 26527186. PMCID: PMC4751994. Epub 20151102.
2. O'Dwyer D, Dickson R, Moore B. The lung microbiome, immunity and the pathogenesis of chronic lung disease. J Immunol. 2016;196(12):4839–47.
3. Bassis C, Erb-Downward J, Dickson R, Freeman C, Schmidt T, Young V, et al. Analysis of the upper respiratory tract microbiotas as the source of the lung and gastric microbiotas in healthy individuals. mBio. 2015;6(2):e00037-15.
4. Dickson R, Huffnagle G. The lung microbiome: new principles for respiratory bacteriology in health and disease. PLoS Pathog. 2015;11(7):e1004923.
5. Gleeson K, Maxwell S, Eggli D. Quantitative aspiration during sleep in normal subjects. Chest. 1997;111(5):1266–72.
6. Huxley E, Viroslav J, Gray W, Pierce A. Pharyngeal aspiration in normal adults and patients with depressed consciousness. Am J Med. 1978;64(4):564–8.
7. Segal L, Clemente J, Tsay J, Koralov S, Keller B, Wu B, et al. Enrichment of the lung microbiome with oral taxa is associated with lung inflammation of a Th17 phenotype. Nat Microbiol. 2016;1:106031.
8. Berg G, Rybakova D, Rischer D, Cernava T, Champomier Verges M, Charles T, et al. Microbiome definition re-visited: old concepts and new challenges. Microbiome. 2020;8:103.
9. Rinninella E, Raoul P, Cintoni M, Franceschi F, Miggiano G, Gasbarrini A, et al. What is the healthy gut microbiota composition? A changing ecosystem across age, environment, diet and diseases. Microorganisms. 2019;7(1):14.
10. Boer C, Radjabzadeh D, Medina-Gomez C, Garmaeva S, Schiphof D, Arp P, et al. Intestinal microbiome composition and its relation to joint pain and inflammation. Nat Commun. 2019;10:4881.
11. Budden K, Gellatly S, Wood D, Cooper M, Morrison M, Hugenholtz P, et al. Emerging pathogenic links between microbiota and the gut-lung axis. Nat Rev Microbiol. 2017;15:55–63.
12. Enaud R, Prevel R, Ciarlo E, Beaufils F, Wieërs G, Guery B, et al. The gut-lung axis in health and respiratory diseases: a place for inter-organ and inter-kingdom crosstalks. Front Cell Infect Microbiol. 2020;10:9. English
13. Ramsheh M, Haldar K, Esteve-Codina A, Purser L, Richardson M, Muller-Quernheim J, et al. Lung microbiome composition and bronchial epithelial gene expression in patients with COPD versus healthy individuals: a bacterial 16S rRNA gene sequencing and host transcriptomic analysis. Lancet Microbe. 2021;2(7):E300–10.
14. Charlson E, Bittinger K, Haas A, Fitzgerald A, Frank I, Yadav A, et al. Topographical continuity of bacterial populations in the healthy human respiratory tract. Am J Respir Crit Care Med. 2011;184(8):957–63.
15. Keely S, Talley N, Hansbro P. Pulmonary-intestinal cross-talk mucosal inflammatory disease. Mucosal Immunol. 2012;5:7–18.
16. Youn H, Lee D, Lee Y, Park J, Yuk S, Yang S, et al. Intranasal administration of live *Lactobacillus* species facilitates protection against influenza virus infections in mice. Antivir Res. 2012;93(1):138–43.
17. Pellaton C, Nutten S, Thierry A, Boudousquie C, Barbier N, Blanchard C, et al. Intragastric and intranasal administration of *Lactobacillus paracasei* NCC2461 modulates allergic airway inflammation in mice. Int J Inflam. 2012;2012:686739.

18. Park M, Ngo V, Kwon Y, Lee Y, Yoo S, Cho Y, et al. *Lactobacillus plantarum* DK119 as a probiotic ocnfers protection against influenza virus by modulating innnate immunity. PLoS One. 2013;8(10):e75368.
19. Jeon H, Kim K, Kim S. Effects of yogurt containing probiotics on respiratory virus infections: influenza H1N1 and SARS-CoV-2. J Dairy Sci. 2023;106(3):1549–61.
20. Lancet T. Global burden of 369 diseases and injuries in 204 countries and territories, 1990–2019: a systematic analysis for the Global Burden of Disease Study 2019. Lancet. 2020;396(10258):1204–22.
21. Strachan DP. Hay fever, hygiene and household size. Br Med J. 1989;299:1259–60.
22. Riedler J, Braun-Farländer C, Eder W, Schreuer M, Waser M, Maisch S, et al. Exposure to farming in early life and development of asthma and allergy: a cross-sectional survey. Lancet. 2001;358(9288):1129–33.
23. Christensen E, Hjelmsø M, Thorsen J, Shah S, Redgwell T, Poulsen C, et al. The developing airway and gut microbiota in early life is influenced by age of older siblings. Microbiome. 2022;10:106.
24. Ball T, Castro-Rodriguez J, Griffith K, Holberg C, Martinez F, Wright A. Siblings, daycare attendance, and the risk of asthma and wheezing during childhood. N Engl J Med. 2000;343:538–43.
25. Ong M, Umetsu D, Mandl K. Consequences of antibiotics and infections in infancy: bugs, drugs, and wheezing. Ann Allergy Asthma Immunol. 2014;112(5):441–5.
26. Donald K, Finlay BB. Early-life interactions between the microbiota and immune system: impact on immune system development and atopic disease. Nat Rev Immunol. 2023; https://doi.org/10.1038/s41577-023-00874-w. PubMed PMID: 37138015. Epub 20230503.
27. Loverdos K, Bellos G, Koklatou L, Vasileiadis I, Giamarellos E, Pecchiari M, et al. Lung microbiome in asthma: current perspectives. J Clin Med. 2019;8(11):1967.
28. Hilty M, Burke C, Pedro H, Cardenas P, Bush A, Bossley C, et al. Disordered microbial communities in asthmatic airways. PLoS One. 2010;5(1):e8578.
29. Marri P, Stern D, Wright A, Billheimer D, Martinez F. Asthma-associated differences in microbial composition of induced sputum. J Allergy Clin Immunol. 2013;131(2):346–52.
30. Huang Y, Nariya S, Harris J, Lynch S, Choy D, Arron J, et al. The airway microbiome in patients with severe asthma: associations with disease features and severity. J Allergy Clin Immunol. 2015;136(4):874–84.
31. Huang Y, Nelson C, Brodie E, DeSantis T, Baek M, Liu J, et al. Airway microbiota and bronchial hyperresponsiveness in patients with suboptimally controlled asthma. J Allergy Clin Immunol. 2011;127(2):372–81.
32. Zhang Q, Cox M, Liang Z, Brinkmann F, Cardenas P, Duff R, et al. Airway microbiota in severe asthma and relationship to asthma severity and phenotypes. PLoS One. 2016;11(4):e0152724.
33. Goleva E, Jackson L, Harris K, Robertson C, Sutherland E, Hall C, et al. The effects of airway mironiome on corticosteroid responsiveness in asthma. Am J Respir Crit Care Med. 2013;188(10):1193–201.
34. Huang E, Inoue T, Leone V, Dalal S, Touw K, Wang Y, et al. Using corticosteroids to reshape the gut microbiome: implication sfor inflammatory bowel diseases. Inflamm Bowel Dis. 2016;21(5):963–72.
35. Franzin M, Stefancic K, Lucafo M, Decorti G, Stocco G. Microbiota and drug response in inflammatory bowel disease. Pathogens. 2021;10(2):211.
36. Hartmann J, Albrich W, Dmitrijeva M, Kahlert C. The effects of corticosteroids on the respiratory microbiome: a systematic review. Fron Med. 2021;8:588584.
37. Tukey M, Lamb C. Bronchial washing, bronchoalveolar lavage, bronchial brush and endobronchial biopsy. In: Ernst A, Herth F, editors. Introduction to bronchoscopy. 2nd ed. Cambridge: Cambridge University; 2017. p. 102–17.
38. García-Vázquez E, Marcos M, Mensa J, de Roux A, Puig J, Font C, et al. Assessment of the usefulness of sputum culture for diagnosis of community-acquired pneumonia using the PORT predictive scoring system. Arch Intern Med. 2004;164(16):1807–11.

39. Grønseth R, Drengenes C, Wiker H, Tangedal S, Xue Y, Husebø G, et al. Protected sampling is preferable in bronchoscopic studies of the airway microbiome. ERJ Open Res. 2017;3(3):00019-2017.
40. Dewhirst F, Chen T, Izard J, Paster B, Tanner A, Yu W, et al. The human oral microbiome. J Bacteriol. 2010;192(19):5002–17.
41. Dickson R, Erb-Downward J, Freeman C, McCloskey L, Beck J, Huffnagle G, et al. Spatial variation in the healthy human lung microbiome and the adapted island model of lung biogeography. Ann Am Thorac Soc. 2015;12(6):821–30.
42. Morris A, Beck J, Schloss P, Campbell T, Crothers K, Curtis J, et al. Comparison of the respiratory microbiome in healthy nonsmokers and smokers. Am J Respir Crit Care Med. 2013;187(10):1067–75.
43. Yatsunenko T, Rey F, Manary M, Trehan I, Dominguez-Bello M, Contreras M, et al. Human gut microbiome viewed across age and geography. Nature. 2012;486:222–7.
44. Natalini J, Singh S, Segal L. The dynamic lung microbiome in health and disease. Nat Rev Microbiol. 2023;21:222–35.
45. Russo C, Colaianni V, Ielo G, Valle M, Spicuzza L, Malaguarnera L. Impact of lung microbiota on COPD. Biomedicine. 2022;10(6):1337.
46. Cuthbertson L, Walker A, Oliver A, Rogers G, Rivett D, Hampton T, et al. Lung function and microbiota diversity in cystic fibrosis. Microbiome. 2020;8:45.
47. Frey D, Bridson C, Dittrich S, Graeber S, Stahl M, Wege S, et al. Changes in the microbiome dominance are associated with declining lung function and fluctuating inflammation in people with cystic fibrosis. Front Microbiol. 2022;13:885822.
48. Yildiz S, Mazel-Sanchez B, Kandasamy M, Manicassamy B, Schmolke M. Influenza A virus infection impacts systemic microbiota dynamics and causes quantitative enteric dysbiosis. Microbiome. 2018;6:9.
49. Sencio V, Barthelemy A, Tavares L, Machado M, Soulard D, Cuinat C, et al. Gut dysbiosis during influenza contributes to pulmonary penumococcal superinfection through altered short-chain fatty acid production. Cell Rep. 2020;30(9):2934–47.
50. Hernández-Terán A, Mejía-Nepomuceno F, Herrera M, Barreto O, Garcia E, Castillejos M, et al. Dysbiosis and structural disruption of the respiratory microbiota in COVID-19 patients with severe and fatal outcomes. Sci Rep. 2021;11:21297.
51. Russell C, Fairfield C, Drake T, Turtle L, Seaton R, Wootton D, et al. Co-infections, secondary infections and antimicrobial use in patients hospitalised with COVID-19 during the first pandemic wave from ISARIC WHO CCP-UK study: a multicentre, prospective cohort study. Lancet Microbe. 2021;8:e354–65.
52. McFarland L. Use of probiotics to correct dysbiosis of normal microbiota following disease or disruptive events: a systematic review. BMJ Open. 2014;4(8):e005047.
53. Hill C, Guarner F, Reid G, Gibson G, Merenstein D, Pot B, et al. The international scientific association for probiotics and prebioics consensus statement on the scope and appropriate use of the term probiotic. Nat Rev Gastroenterol Hepatol. 2014;11:506–14.
54. Harata G, He F, Hiruta N, Kawase M, Kybota A, Hiramatsu M, et al. Intranasal administraton of *Lactobacillus rhamnosus* GG protects mice from H1N1 infuenza virus infection by regulating respiratory immue responses. Lett Appl Microbiol. 2010;50(6):597–602.
55. Tomosada Y, Chiba E, Zelaya H, Takahashi T, Tsukida K, Kitazawa H, et al. Nasally administered *Lactobacillus rhamnosus* strains differentially modulate respiratory antiviral immune responses and induce protection against respiratory syncytial virus infection. BMC Immunol. 2013;14:40.
56. Zelaya H, Tada A, Vizoso-Pinto M, Salva S, Kanmani P, Aguero G, et al. Nasal priming with immunobiotic *Lactobacillus rhamnosus* modulates inflammation-coagulation interactions and reduces influenza virus-associated pulmonary damage. Inflamm Res. 2015;64:589–602.
57. Jung Y, Lee Y, Ngo V, Cho Y, Ko E, Hong S, et al. Heat-killed *Lactobacillus casei* confers broad protection against influenza A virus primary infection and develops heterosubtypic immunity against future secondary infection. Sci Rep. 2017;7:17360.

58. Izumo T, Maekawa T, Ida M, Noguchi A, Kitagawa Y, Shibata H, et al. Effect of intranasal administration of *Lactobacillus pentosus* S-PT84 on influenza virus infection in mice. Int Immunopharmacol. 2010;10(9):1101–6.
59. Park M, Ngo V, Kwon Y, Lee Y, Yoo S, Cho Y, et al. *Lactobacillus plantarum* DK119 as a probiotic confers protection against influenza virus by modulating innnate immunity. PLoS One. 2013;8(10):e75368.
60. Percopo C, Rice T, Brenner T, Dyer K, Luo J, Kanakabandi K, et al. Immunobiotic *Lactobacillus* administered post-exposure averts the lethal sequelae of respiratory virus infection. Antivir Res. 2015;121:109–19.
61. Groeger D, Schiavi E, Grant R, Kurnik-Lucka M, Michalovich D, Williamson R, et al. Intranasal *Bifidobacterium longum* protects against viral-induced lung inflammation and injury in a murine model of lethal influenza infection. EBioMedicine. 2020;60:102981.
62. Medina M, Villena J, Salva S, Vintiñi E, Langella P, Alvarez S. Nasal administration of *Lactococcus lactis* improves local and systemic immune responses against *Streptococcus pneumoniae*. Microbiol Immunol. 2008;52(8):399–409.
63. Kim H, Arellano K, Park H, Todorov S, Kim B, Kang H, et al. Assessment of the safety and anti-inflammatory effects of three *Bacillus* strains in the respiratory tract. Environ Microbiol. 2021;23(6):3077–98.
64. Tonetti F, Islam M, Vizoso-Pinto M, Takahashi H, Kitazawa H, Villena J. Nasal priming with immunobiotic lactobacilli improves the adaptive immune response against influenza virus. Int Immunopharmacol. 2020;78:106115.
65. Tonetti F, Islam M, Vizoso-Pinto M, Salva S, Kanmani P, Aguero G, et al. Nasal priming with immunobiotic Lactobacillus rhamnosus modulates inflammation-coagulation interactions and reduces influenza virus-associated pulmonary damage. Inflamm Res. 2015;64:589–602.
66. Islam M, Albarracin L, Tomokiyo M, Valdez J, Sacur J, Vizoso-Pinto M, et al. Immunobiotic lactobacilli improve resistance of respiratory epithelial cells to SARS-CoV-2 infection. Pathogens. 2021;10(9):1197.
67. Kolling Y, Salva S, Villena J, Alvarez S. Are the immunomodulatory properties of *Lactobacillus rhamnosus* CRL1505 peptidoglycan common for all Lactobacilli during respiratory infection in malnourished mice? PLoS One. 2018;13(3):e0194034.
68. Kolling Y, Salva S, Villena J, Marranzino G, Alvarez S. Non-viable immunobiotic Lactobacillus rhamnosus CRL1505 and its peptidoglycan improve systemic and respiratory innate immune response during recovery of immunocompromised-malnourished mice. Int Immunopharmacol. 2015;25(2):474–84.
69. Cooper M, Fehniger T, Ponnappan A, Mehta V, Wewers M, Caligiuri M. Interleukin-1β costimulates interferon-γ production by human natural killer cells. Eur J Immunol. 2001;31(3):792–801.
70. Renegar K, Small P Jr, Boykins L, Wright P. Role of IgA versus IgG in the control of influenza viral infection in the murine respiratory tract. J Immunol. 2004;173(3):1978–86.
71. Oliveria-Nascimento L, Massari P, Wetzler L. The role of TLR2 in infection and immunity. Front Immunol. 2012;3:79.
72. Ivory K, Wilson A, Sankaran P, Westwood M, McCarville J, Brockwell C, et al. Oral delivery of a probiotic induced changes at the nasal mucosa of seasonal allergic rhinitis subjects after local allergen challenge: a randomised clinical trial. PLoS One. 2013;8(11):e78650.
73. Zeng J, Wang C, Zhang F, Qi F, Wang S, Ma S, et al. Effect of probiotics on the incidence of ventilator-associated pneumonia in critically ill patients: a randomised controlled multicentre trial. Intensive Care Med. 2016;42:1018–28.
74. Endam L, Alromaih S, Gonzalez E, Madrenas J, Cousineau B, Renteria A, et al. Intranasal application of *Lactococcus lactis w136* is safe in chronic rhinosinusitis patients with previous sinus surgery. Front Cell Infect Microbiol. 2020;10:440.
75. Southern California Evidence-based Practice Center. Safety of probiotics to reducee risk and prevent or treat disease. In: U.S. Department of Health and Human Services, editor. Evidence report/technology assessment no. 200, AHRQ Publication No. 11-E007. Rockville: Agency for Healthcare Research and Quality; 2011.

76. Tran D, Tran T, Phung T, Bui H, Nguyen P, Vu T, et al. Nasal-spraying *Bacillus* spores as an effective symptomatic treatment for children with acute respiratory syncytial virus infection. Sci Rep. 2022;12:12402.
77. De Boeck I, van den Broek M, Allonsius C, Spacova I, Wittouck S, Martens K, et al. Lactobacilli have a niche in the human nose. Cell Rep. 2020;31(8):107674.
78. Spacova I, De Boeck I, Cauwenberghs E, Delanghe L, Bron P, Henkens T, et al. Development of a live biotherapeutic throat spray with lactobacilli targeting respiratory viral infections. Microb Biotechnol. 2022;16:99–115.
79. EFSA BIOHAZ Panel, Koutsoumanis K, Allende A, Alvarez-Ordonez A, Bolton D, Bover-Cid S, et al. Updated list of QPS-recommended microorganisms for safety risk assessments carried out by EFSA. 2023.
80. De Boeck I, Cauwenberghs E, Spacova I, Gehrmann T, Eliers T, Delanghe L, et al. Randomised, double-blind, placebo-controlled trial of a throat spray with selected lactobacilli in COVID-19 outpatients. Microbiol Spectr. 2022;10(5):e01682-22.
81. Lee C, Choi Y, Seo S, Kim S, Kim I, Kim S, et al. Addition of probiotics to antibiotics improves the clinical course of pneumonia in young people without comorbidities: a randomised controlled trial. Sci Rep. 2021;11:926.
82. Chong H, Yusoff N, Hor Y, Lew L, Jaafar M, Choi S, et al. *Lactobacillus plantarum* DR7 improved upper respiratory tract infections via enhancing immune and inflammatory parameters: a randomised, double-blind, placebo-controlled study. J Dairy Sci. 2019;102(6):4783–97.
83. Johnstone J, Meade M, Lauzier F, Marshall J, Duan E, Dionne J, et al. Effect of probiotics on incident ventilator associated pneumonia in critically ill patients: a randomised clinical trial. JAMA. 2021;326(11):1024–33.
84. Fangous M, Alexandre Y, Hymery N, Gouriou S, Arzur D, Le Blay G, et al. Lactobacilli intra-tracheal administration protects from *Pseudomonas aeruginosa* pulmonary infection in mice – a proof of concept. Benef Microbes. 2019;10(8):893–900.
85. Mårtensson A, Nordström F, Cervin-Hoberg C, Lindstedt M, Skakellariou C, Cervin A, et al. Nasal administration of a probiotic assemblage in allergic rhinitis: a randomised placebo-controlled crossover trial. Clin Exp Allergy. 2022;52(6):774–83.
86. Spacova I, Fremau P, Pollaris L, Vanoirbeek J, Ceuppens J, Seys S, et al. Intranasal administration of probiotic *Lactobacillus rhamnosus* GG prevents birch pollen-induced allergic asthma in a murine model. Allergy. 2019;74:100–10.
87. Morales A, Carvajal P, Silva N, Hernandez M, Godoy C, Rodriguez G, et al. Clinical effects of *Lactobacillus rhamnosus* in non-surgical treatment of chronic periodontitis: a randomised placebo-controlled trial with 1-year follow-up. J Periodontol. 2016;87(8):944–52.
88. Lee X, Vergara C, Lozano C. Severity of *Candida*-associated denture stomatitis is improved in institutionalised elders who consume *Lactobacillus rhamnosus* SP1. Aust Dent J. 2019;64(3):229–36.
89. Henriksson R, Franzén L, Sandström K, Nordin A, Arevärn M, Grahn E. Effects of active addition of bacterial cultures in fermented milk to patients with chronic bowel discomfort following irradiation. Support Care Cancer. 1995;3:81–3.
90. Huang S, Vignolles M, Chen X, Le Loir Y, Jan G, Schuck P, et al. Spray drying of probiotics and other food-grade bacteria: a review. Trends Food Sci Technol. 2017;63:1–17.
91. Fu N, Huang S, Xiao J, Chen X. Chapter 6: Producing powders containing active dry probiotics with the aid of spray drying. Adv Food Nutr Res. 2018;85:211–62.
92. Santivarangkna C, Higl B, Foerst P. Protection mechanisms of sugars during different stages of preparation process of dried lactic acid starter cultures. Food Microbiol. 2008;25(3):439–41.
93. Su Y, Zheng X, Zhao Q, Fu N, Xiong H, Wu W, et al. Spray drying of *Lactobacillus rhamnosus* GG with calcium-containing protectant for enhanced viability. Powder Technol. 2019;358:87–94.
94. Prasad J, McJarrow P, Gopal P. Heat and osmotic stress responses of probiotic Lactobacillus rhamnosus HN001 (DR20) in relation to viability after drying. Appl Environ Microbiol. 2003;69(2):917–25.

95. Iaconelli C, Lemetais G, Kechaou N, Chain F, Bermudez-Humaran L, Langella P, et al. Drying process strongly affects probiotics viability and functionalities. J Biotechnol. 2015;214:17–26.
96. Zhang Y, Lin J, Zhong Q. Effects of media, heat adaptation, and outlet temperature on the survival of *Lactobacillus salivarius* NRRL B-30514 after spray drying and subsequent storage. LWT. 2016;74:441–7.
97. Barbosa J, Almeida D, Machado D, Sousa S, Freitas A, Andrade J, et al. Spray-drying encapsulation of the live biotherapeutic candidate *Akkermansia muciniphila* DSM 22959 to survive aerobic storage. Pharmaceuticals. 2022;15(5):628.
98. Behboudi-Jobbehdar S, Soukoulis C, Yonekura L, Fisk I. Optimisation of spray-drying process conditions for the production of maximally viable microencapsulated *L. acidophilus* NCIMB 701748. Dry Technol. 2013;31(11):1274–83.
99. Perdana J, Bereschenko L, Fox M, Kuperus J, Kleerebezem M, Boom R, et al. Dehydration and thermal inactivation of *Lactobacillus planatarum* WCFS1: comparing single droplet drying to spray and freeze drying. Food Res Int. 2013;54(2):1351–9.
100. Paez R, Lavari L, Vinderola G, Audero G, Cuatrin A, Zaritzky N, et al. Effect of heat treatment and spray drying on lactobacilli viability and resistance to simulated gastrointestinal digestion. Food Res Int. 2012;48(2):748–54.
101. Kojima M, Suda S, Hotta S, Hamada K, Suganuma A. Necessity of calcium ion for cell division in *Lactobacillus bifidus*. J Bacteriol. 1970;104(2):1010–3.
102. Yu A, Wei L, Marco M. Calcium determines *Lactiplantibacillus plantarum* intraspecies competitive fitness. Appl Environ Microbiol. 2022;88(15):e00666-22.
103. Huang S, Chen X. Signficiant effect of Ca2+ on improving the heat resistance of lactic acid bacteria. FEMS Microbiol Lett. 2013;344(1):31–8.
104. Yang Y, Huang S, Wang J, Jan G, Jeantet R, Chen X. Mg2+ improves the thermotolerance of probiotic *Lactobacillus rhamnosus* GG, *Lactobacillus casei* Zhang, *Lactobacillus plantarum* P-8. Lett Appl Microbiol. 2017;64(4):283–8.
105. Jokicevic K, Kiekens S, Byl E, De Boeck I, Cauwenberghs E, Lebeer S, et al. Probiotic nasal spray development by spray drying. Eur J Pharm Biopharm. 2021;159:211–20.
106. Roos Y, Karel M. Plasticising effect of water on thermal behaviour and crystallisation of amorphous food models. J Food Sci. 1991;56(1):38–43.
107. Simperler A, Kornherr A, Chopra R, Bonnet A, Jones W, Motherwell S, et al. Glass transition temperature of glucose, sucrose and trehalose: an experimental and in silico study. J Phys Chem. 2006;110:19678–84.
108. Zillen D, Beugeling M, Hinrichs W, Frijlink H, Grasmeijer F. Natural and bioinspired excipients for dry powder inhalation formulations. Curr Opin Colloid Interface Sci. 2021;56:101497.
109. Rajab S, Tabandeh F, Shahraky M, Alahyaribeik S. The effect of *lactobacillus* cell size on its probiotic characteristics. Anaerobe. 2020;62:102103.

Respiratory Delivery of Bacteriophages for the Treatment of Lung Infections

Alex Seungyeon Byun, Hak-Kim Chan, and Philip Chi Lip Kwok

Abstract Bacteriophages (phages) are highly specialised in targeting bacteria down to the strain level. Rising concerns regarding antibiotic resistance have fueled research on using bacteriophages to combat this significant threat. Numerous in vitro and in vivo studies demonstrating the efficacy of phages have been completed to strengthen their use as potential therapeutics against muti-drug resistant (MDR) pathogens. Current evidence supports the use of inhaled phages in respiratory infections such as pneumonia. However, compared to the oral and intravenous routes, the inhalation route has been underexplored, especially in randomised controlled trials. This review explores the current literature on inhaled phage therapy and the efforts in formulating safe, efficacious phages for inhalation.

Keywords Bacteriophages · Phages · Bacteria · Infection · Multi-drug resistance

8.1 Introduction

Bacteriophages are ubiquitous and the most abundant life form, with about 10^{31} phage particles on Earth [1]. As a natural fighter against bacteria, there are many species discovered and to be discovered. The therapeutic potential of this "obligate intracellular parasite of bacteria" was proposed in 1917 by the French-Canadian microbiologist, Félix d'Hérelle [2]. However, its development in the West was interrupted by World War II and the advent of antibiotics. On the other hand, Eastern Europe continued studying bacteriophages during that period.

A. S. Byun · H.-K. Chan · P. C. L. Kwok
Advanced Drug Delivery Group, Sydney Pharmacy School, Faculty of Medicine and Health, The University of Sydney, Camperdown, NSW, Australia
e-mail: philip.kwok@sydney.edu.au

J. K. W. Lam, P. C. L. Kwok (eds.), *Respiratory Delivery of Biologics, Nucleic Acids, and Vaccines*, AAPS Introductions in the Pharmaceutical Sciences 8,
https://doi.org/10.1007/978-3-031-47567-2_8

Antimicrobial resistance (AMR) threatens our health and well-being. It has been forecasted to be responsible for ten million deaths by 2050, making it more lethal than cancer [3]. The World Health Organisation has indicated AMR to be a major health crisis that we face in the twenty-first century. We are challenged by the growing drug resistance against the ESKAPE bacteria (*Enterococcus faecium*, *Staphylococcus aureus*, *Klebsiella pneumoniae*, *Acinetobacter baumannii*, *Pseudomonas aeruginosa*, *Enterobacter* spp.). The lack of incentive for the research and development of new antibiotics further fuels the threat of AMR. Consequently, there is increasing interest in combating AMR with phages.

Phages are viruses that specifically infect bacteria and can be classified into two types according to how they replicate, namely, lysogenic, or lytic phages. Lysogenic phages incorporate their viral gene into the host genome and the transmission of viral genetic material to daughter bacterial cells. On the other hand, lytic phages hijack the bacteria to produce phage components for phage replication and assembly within the bacterial host. The new phages are released when the cell lyses (Fig. 1). Therapeutically, strictly lytic phages are utilised due to their rapid replication and lysis of bacteria. The narrow spectrum of infectivity allows phages to target unwanted pathogens whilst avoiding beneficial microbes [4, 5] and also enables localised treatment [6] in contrast to antibiotics. As phages are living organisms, they have the capacity to co-evolve with their bacterial host, possessing the potential to tackle AMR [7]. Studies have shown promising therapeutic results against MDR

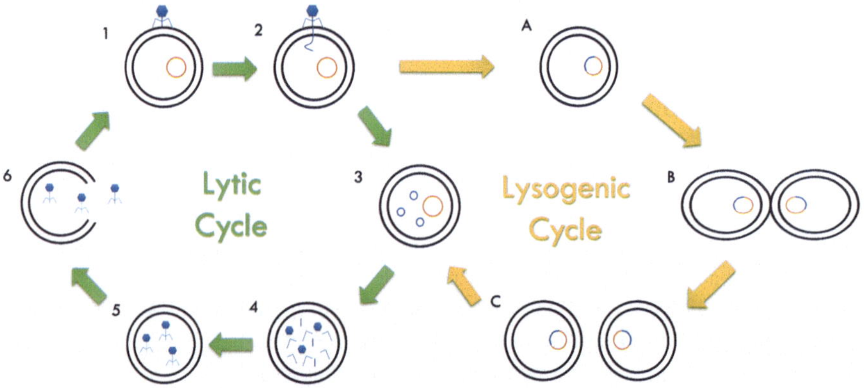

Fig. 1 Schematic of the replication of bacteriophages. Bacteriophages can replicate by either the lytic or the lysogenic cycle depending on their species. Lytic cycle: (1) Phages recognise their specific host and dock themselves to the bacterial membrane. (2) The attached phage releases its DNA into bacterial cells. (3) Phage DNA replicates and circularises but remains separate from the host DNA. (4) Phage proteins are made from the replicated DNA. (5) Phages parts assemble. (6) As phages fill the host cell, it lyses thereby releasing the phages. The lysogenic cycle takes a slightly different course from the lytic cycle. (A) Once the phage releases its DNA into the bacterial cell, the phage DNA is integrated into the host genome. The genetic material of phages is referred to as prophages. (B) Bacterial cell divides, replicating the prophage. (C) The host cells continue to replicate with the integrated prophage. When the bacterial cell experiences stress, it can excise the prophage (Step 3) and undergo the lytic cycle to replicate phages

bacterial strains in both animals [6, 8–11] and clinical case studies [12–14]. Before phage therapy can be widely used for treating infections, its therapeutic efficacy, immunogenicity, and its application as different formulations need to be investigated.

8.2 Inhaled Phages Against MDR Pathogens

Phages have demonstrated positive outcomes in both animal and clinical studies. Therapeutically, lytic phages are preferred as the cycle of infection, replication, and lysis is quicker than that of lysogenic phages. Its specificity and ability to multiply are desirable as it may lower the risk of adverse events and is able to provide a constant dose supply, meaning it may penetrate biofilms [15]. Recent successful case studies on treating clinical MDR infections are promising [12, 13]. Given the rising AMR and lack of new antibiotic development, it is worth pursuing and realising the therapeutic potential of phages. Inhaled phage delivery would be more efficient than other routes of administration for respiratory infections because it targets pathogens in the lungs directly, elicits faster action, and is non-invasive.

8.2.1 Staphylococcus aureus

Methicillin-resistant *S. aureus* (MRSA) commonly causes nosocomial infections and is resistant to first-line antibiotics. They are responsible for ventilator-associated pneumonia (VAP), hence treatment with a phage cocktail has been investigated [16, 17]. Phages 2003, 2002, 3A, and K at a total concentration of 1.5×10^{10} plaque forming units (PFU)/mL were nebulised (vibrating mesh) at 2, 12, 24, 48, 72 hours post-inoculation of the MRSA mouse model and compared with intravenous (IV) delivery of the phage cocktail and IV linezolid [16]. The phage cocktail alone was successful in improving the survival of the mice by 50% compared to control (0%) and IV linezolid (38%) and was solely localised in the lungs, indicating no association with systemic side effects [16]. However, when administered together with an IV phage cocktail, the survival rate dramatically improved to 91% [16]. The nebulised phages only rescued 50% of the mice, possibly due to poor penetration into the deep lung parenchyma as well as the infection spreading into the spleen, indicating bacteraemia [16]. Phage therapy could also be utilised as prophylaxis for VAP [17]. A single high dose of the nebulised phage cocktail (2003, 2002, 3A, K) 6 hours prior to infection resulted in 60% survival for 3×10^{10} PFU and 70% for 3×10^{11} PFU compared to untreated mice, none of which survived [17]. The bacterial load in the lungs of the surviving mice was approximately 500 times lower than that in the untreated and dead mice [17]. The authors concluded that these studies could be the foundations for future clinical trials as currently there are no randomised controlled trials (RCT) examining inhaled phage therapy against *S. aureus*.

8.2.2 Klebsiella pneumoniae

K. pneumoniae is an opportunistic pathogen commonly found to cause community- and hospital-acquired pneumonia. Like many other pathogenic species, many strains are becoming increasingly multi-drug resistant, including against carbapenems [18]. Intranasal administration of phages pKp11 and pKp383 resulted in 100% survival and boasted a greater efficiency in suppressing the bacterial load in mice infected with a clinical isolate, *K. pneumoniae* SY1 compared to treatment with either pKp11 or pKp383 alone [19]. In a mouse pneumonia model, a single intranasal administration of monophage 2 hours post-inoculation of *K. pneumonia* resulted in 80% survival compared to the PBS control [20]. The bacterial load detected in phage-treated mice with sublethal pneumonia was 2-logs lower, the observed lung consolidation was localised, and lung tissues were predominantly healthy compared to the control. Similarly, another study which investigated the mouse pneumonia model observed no inflammatory cell exudate in the bronchioles, focussed areas of necrosis as opposed to diffuse areas in histopathological lesions, and no perivascular/bronchial cuffing in mice treated with phages compared to control [21]. Overall, there was a significant drop in bacterial load 48 hours post-infection and a complete elimination was achieved at 96 hours post-infection.

8.2.3 Acinetobacter baumanni

A. baumannii is also another opportunistic pathogen that is often associated with hospital-acquired pneumonia and carbapenem-resistant *A. baumannii* (CRAB) is a growing public health problem. An in vivo study demonstrated better survival for the intranasal PBAB phage cocktail-treated (PBAB08, PBAB25) mice compared to the control at day 4 post-infection (60% vs 20% survival) [8]. A 2-log reduction in bacterial load in the infected lungs of mice was noted on days 3–4 post-infection. However, a complete elimination of the bacteria was not achieved. Similarly, treatment with intranasal anti-baumannii phage Bϕ-R2096 resulted in improved survival 16 and 48 hours post-infection compared to control in mice [10]. The authors also reported that the phage-only treatment did not show any tissue damage and no serious adverse events were noted. Interestingly, Bϕ-R2096 was able to completely clear the CRAB infection in the lungs, unlike the PBAB cocktail. Although this indicates the therapeutic efficacy of phages, it also highlights the complexity of selecting appropriate phages for therapeutic use. The differing extent of survival rates could depend on several factors, such as different strains of CRAB, different mouse types, and dosing regimens.

8.2.4 Pseudomonas aeruginosa

P. aeruginosa frequently infects cystic fibrosis patients and causes pneumonia. Mouse inhalation model that realistically reflects chronic lung infection caused by *P. aeruginosa* demonstrated a 3-log reduction in bacteria colony forming units (CFU) 24 hours post-intranasal phage PELP20 treatment [11]. Ultimately, a complete clearance of infection was observed in 70% of the mice and the remaining 30% had significant CFU reduction 6 days post-infection [11]. Nebulised phage cocktail therapy rapidly controlled ventilation-associated pneumonia in a recent porcine model that closely resembled focal pneumonia in humans [22]. The rapid 1.5 log reduction in bacterial load may be due to the higher-than-expected phage titre of 10^7–10^9 PFU/mL that was observed in the broncoalveolar lavage (BAL) and in consolidated pneumonia. This suggests that phages can replicate on-site and penetrate deep areas of infection where there may be limited air passage. Furthermore, no phages were detected in the serum or urine, again highlighting its localised action.

Success of phage therapy against *P. aeruginosa* was also observed in humans, but only through intravenous administration. A case study of a cystic fibrosis patient with pneumonia caused by MDR *P. aeruginosa* was administered with an intravenous (IV) cocktail of four lytic phages at a dose of 4×10^9 PFU/5 mL every 6 hours for 8 weeks [13]. This patient remained on azithromycin and piperacillin-tazobactam prior to and throughout the phage course and had intermittent use of ciprofloxacin and doripenem during phage therapy. By the end of the phage course, the infection was fully resolved without adverse reactions and with no recurrence within 100 days. Another case study demonstrated the success and safety of phage monotherapy in a lung transplant patient with *P. aeruginosa* pneumonia [12]. However, these are isolated success reports in which phages were used under compassionate grounds. They only serve as non-systematic evidence to encourage further clinical trials. Furthermore, phage therapy in humans has been limited to oral, IV, and topical routes. The inhalation route is yet to be investigated, despite the countless number of studies that have reported its efficacy. Currently, there are three registered RCTs that evaluate the safety and/or efficacy of inhaled phages against *P. aeruginosa* (clinical trial identifier: NCT05616221, NCT04596319, NCT05010577) [23–25]. The scarcity of clinical evidence on inhaled phages may be due to the lack of robust safety data and hesitation in administering phages that were not produced by Good Manufacturing Practice (GMP). The PhagoBurn trial in 2019 [26] was the only double-blind RCT investigating the efficacy of phages that were produced by GMP and gained approval from health regulatory agencies in France, Belgium, and Switzerland.

8.2.5 Mycobacterium tuberculosis

M. tuberculosis is an intracellular bacterium that causes tuberculosis and can quickly mutate to become resistant to antibiotics and typically targets lung alveolar macrophages [27]. As there are numerous natural phages, there would be ones that specifically target *M. tuberculosis*. An example is phage D29, which was used as prophylaxis against the pathogen [28]. A dose of 1 plaque forming unit (PFU)/alveolus was delivered using nose-only inhalation to mice via a vibrating mesh nebuliser 30 minutes prior to *M. tuberculosis* H37Rv infection [28]. Prophylaxis treatment with nebulised D29 demonstrated a significant reduction in bacterial load 24 hours post-infection (approximately one-third of the untreated mice) and continuously maintained a low bacterial burden 3 weeks post-infection [28]. Although the infections were not caused by *M. tuberculosis*, phage therapy has been compassionately used to treat 20 patients against infections caused by the non-tuberculosis-causing *Mycobacterium* spp. [29]. The outcome was inconclusive, with one patient completely resolved of infection after 1 year of IV phage treatment, whilst one patient stopped treatment after 6 months due to a lack of response [29]. Another patient was also completely cleared of the infection, but only after receiving IV phages for 10 months plus two direct bronchoscopic administrations. One patient was free from infection after IV treatment for 1 year but is still continuing therapy [29]. Nineteen out of twenty patients initially received IV phage therapy. Five of them changed to aerosolised phage treatment that is still ongoing at the time of writing [29]. The authors did not specify how they aerosolised the phages. Treatment of *Mycobacterium* is difficult as phages would need to enter, and remain viable after entering, the host cell in order to infect the bacteria inside. Although clinical evidence is limited so far, utilising phages to treat intracellular bacterial infections is promising. More research in this area is needed.

8.3 Phage Synergy

Phages provide a promising treatment against AMR as some phages have synergy with antibiotics. The synergy manifests through enhancing the susceptibility of the bacteria to antibiotics, increasing the lytic activity of the phages, or a higher reduction of the bacterial load. It may be possible to delay or even avoid the development of phage-resistant bacteria [30]. The task force convened by the Antibacterial Resistance Leadership Group recommended that phage therapy should be used in conjunction with conventional antibiotics [14], increasing the need to investigate the synergistic effects of phages and antibiotics. Studies have shown that synergy depends on the species of the phage.

Ciprofloxacin is often used to treat respiratory infections and works synergistically with phages. Lin et al. [31] reported that anti-Pseudomonas phage PEV20 and ciprofloxacin displayed a synergistic effect which was maintained even after jet nebulisation. The mechanism behind the synergy was suggested to be due to PEV20 changing the activity of the efflux pump on the bacteria to increase the intracellular concentration of ciprofloxacin which then inhibits bacterial DNA replication [31] (Fig. 2). Lin et al. [32] also demonstrated that the synergy was maintained in a mouse lung infection model when co-spray dried PEV20 with ciprofloxacin was delivered intratracheally, which reduced $CD8^{+}T$ cells, B cells, and lung monocytes/

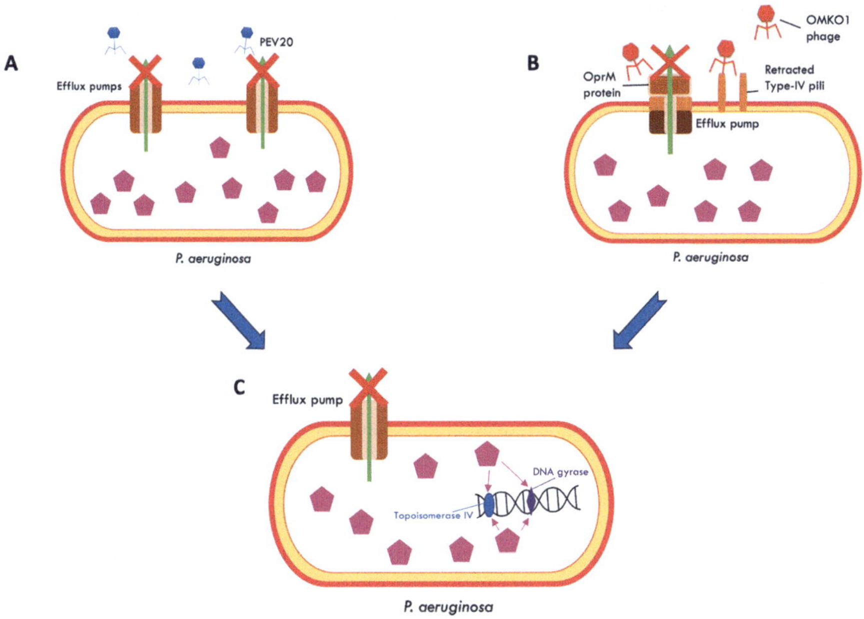

Fig. 2 Proposed mechanism of action of phage-antibiotic synergy. (A) Lin et al. [31] have suggested that phage PEV20 may induce changes in the efflux pumps of *P. aeruginosa*, increasing the intracellular concentration of ciprofloxacin (represented by pink pentagons). Further clarification into how PEV20 may interact with the efflux pumps has not been explored. (B) Gurney et al. [33] proposed that phage OMKO1 may bind to the surface of *P. aeruginosa* via Type-IV pili, causing the pili to retract for an easier interaction with the OprM protein of the efflux pump. Ultimately this will retain ciprofloxacin concentration in the bacteria. This mechanism is also applicable to tetracycline. (C) The high intracellular concentration of ciprofloxacin works to inhibit DNA gyrase and topoisomerase IV to disrupt DNA synthesis, leading to cell death

macrophages. However, the decreased inflammatory response might be due to the rapid control of bacterial infection by ciprofloxacin-PEV20 and not necessarily due to active anti-inflammation by the combination. Other *P. aeruginosa* targeting phages showed synergy with antibiotics by reducing antibiotic resistance but it should be noted that it was not delivered via the respiratory route [30, 33–35]. In fact, the two studies by Lin et al. [31, 32] are the only published studies so far that examine inhaled phage antibiotic synergy (PAS) [36]. The same phage cocktail as that used in PhagoBurn (PP1131) reduced the bacterial load of the resistant *P. aeruginosa* CHA strain in an in vitro fibrin-clot model 6 hours post-administration, but not the P7 bacterial strain [30]. However, bacterial regrowth was observed beyond 6 hours post-administration, indicating phage-resistance development. In vivo, endocarditis was induced in rats by inoculation with *P. aeruginosa* CHA and treated with phage and ciprofloxacin intravenously. The combination resulted in negative vegetation cultures in 7 out of 11 rats 6 hours after treatment whereas ciprofloxacin-alone and phage-alone had no rats with negative cultures. Unlike in vitro, phage resistance was not detected in vivo where co-therapy with ciprofloxacin was applied [30]. Phage resistance in vivo is not commonly observed due to a fitness trade-off or clearance by the immune system [37, 38], which may be the case in the study by Oechslin et al. as neutropenic mice were not used. Further investigation has identified the characteristics of phage-resistant bacterial strains. The *P. aeruginosa* 24/2 strain had a disrupted *pilT* gene which produces ATPase for type IV pilus contraction, whereas the *P. aeruginosa* 19/2 strain had a truncated lipopolysaccharide (LPS) that resulted in reduced virulence. Type IV pili and LPS are also phage receptors in which disruptions have resulted in phage resistance and hence the regrowth of *P. aeruginosa* 6 hours post-treatment [30]. The regrowth of *P. aeruginosa* in the presence of a cocktail of 12 different phages implied that the therapeutic dose of the individual phages was suboptimal, resulting in ineffective clearance of the bacteria [30]. It could also be due to the various phages not complementing each other [30]. Cocktails should be designed strategically, considering the host range and targeting receptors on the bacteria to be effective in reducing or eliminating phage-resistance development [39].

Similarly, co-culturing *P. aeruginosa* with phage OMKO1 resulted in a >60% reduction in antibiotic resistance to erythromycin, tetracycline, and ciprofloxacin and also displayed "phage steering" where phage-resistance is traded off with antibiotic-resistance [33]. OMKO1 increased bacterial susceptibility to ciprofloxacin and tetracycline in addition to preventing antibiotic resistance evolution (Fig. 2). Ceftazidime was also used with OMKO1 to treat prosthetic vascular graft infections in a patient [34]. The fact that phage OMKO1 works synergistically with anti-pneumonia drugs, such as ciprofloxacin and ceftazidime, makes it a good candidate for the respiratory route. Direct delivery of OMKO1 and either of these antibiotics to the lung may exert a more efficient response coupled with PAS to quickly subdue the bacterial infection. This may also be applicable to the phage cocktail PP1131. The exploration of the respiratory route with PP1131 would be a big step in inhaled phage therapy as the phage cocktail was manufactured by GMP. Furthermore, Roach et al. [38] showed that the immune system plays a major role in delivering effective inhaled phage therapy. Phage therapy is multidimensional. The literature indicates that phages could be used to potentiate antibiotics to once again target bacterial infections which were once multi-drug resistant.

8.4 Immune Responses Induced by Respiratory-Delivered Phages

Cytokine storms can be observed in both infectious and non-infectious diseases and can also result from therapeutic interventions [40, 41]. In the event of a cytokine storm during severe lung infections, local inflammation overflows into the systemic circulation and causes sepsis [40]. As the therapeutic phages are lytic, the possibility of an uncontrolled surge in endotoxin release following bacterial lysis and the consequent pro-inflammatory response cannot be dismissed. Even when phages do not cause a cytokine storm, they may elevate, suppress, or modulate both pro-inflammatory and anti-inflammatory cytokines. To fully utilise the therapeutic potential of phages, we must investigate the nature of the immune responses that phages induce in vivo.

There is currently a limited number of studies on the immune response of phages targeting *P. aeruginosa* via the respiratory route and their results are inconclusive (Table 1). Intranasal delivery of a phage cocktail targeting *P. aeruginosa* reduced the bacterial load and lowered pro-inflammatory cytokines, including IL-6, TNF-α, keratinocyte chemoattractant, and IL-12p70 in mice [42]. Similarly, in the treatment of pneumonia colonised with *K. pneumoniae*, intranasally delivered phage 1513 significantly reduced TNF-α and IL-6 levels than the untreated mice [20]. However, the changes in the immune response were only observed in the presence of the bacteria so it is unclear whether the cytokine reduction was due to phage activity alone or due to the bacterial load reduction.

Dufour et al. [43] have not only compared the inflammatory response of phages to uninfected controls in healthy mice but also to antibiotics in acute *E. coli* pneumonia in mice. In uninfected mice, both phage 536_P1 and LM33_P1 did not significantly change the complete blood count compared to the control, except that LM33_P1 reduced the lymphocyte count at 7 hours post-administration but that returned to normal 10 hours after that. Phage 536_P1 elevated IFN-y, IL-12, MCP-1 and keratinocyte chemoattractant at 17 hours post-administration, whereas LM33_P1 showed no difference in the control at any time point. Phage 536_P1 also decreased IL-6, IL-1β, TNF-α, keratinocyte chemoattractant, and MCP-1 to similar levels to that achieved by ceftriaxone, a commonly used antibiotic for pneumonia.

Intratracheal delivery of *P. aeruginosa* PAO1 and filamentous phage Pf4 resulted in a reduction of CXCL1, IL-17, IL-1β, IL-17 and IL-10. However, the pro-inflammatory cytokine, IL-12p40, was significantly higher at 48 hours post-administration compared to PAO1-infection alone [44]. In contrast, intranasal administration of a phage cocktail did not affect serum cytokines such as IL-1a, IL-1b, IL-2, IL-4, IL-6, IL-10, IL-12, IL-17A, TNF-a and G-CSF [8]. Similarly, intranasal instillation of phage PAK_P1 showed no significant increase in cytokine production compared to the control except for IFN-y and TNF-α [38]. Interestingly, PAK_P1 displayed immunophage synergy where it relied on the host's immune system to effectively resolve the bacterial infection. This response is yet to be observed in other species of phages. There is currently no literature investigating the adaptive responses of phages in the lungs.

8.5 Formulations and Stability

Stability is of utmost importance for any drug candidate, whether it is a small molecule or biologic. Instability affects the delivery of the correct dose and bioavailability and can form degradants. The PhagoBurn trial failed to show a clinical

Table 1 Immunological effects of respiratory delivered phages in animal lung infection models

Phage	Pathogen	Route of administration/ animal	Dose	Immunological effects	Reference
Phage cocktail 1 (contains: 24, 25, 7)	*P. aeruginosa* PAO1	Intranasal/ BALB/c mice	1.24×10^9 PFU	Reduced IL-10, IL-6, TNF-α, keratinocyte chemoattractant, IL-12p70 in the bronchoalveolar lavage fluid Phage treatment significantly reduced neutrophils compared to untreated mice	[42]
KP 1513	*K. pneumoniae* KP 1513	Intranasal/ Swiss-Webster mice	2×10^9 PFU	Reduced IL-6 and TNF-α in lung tissue compared to untreated mice	[20]
536_P1, LM33_P1	*E. coli* 536, *E. coli* LM33	Intranasal/ BALB/c mice	10^7 CFU for phage 536 5×10^7 CFU for phage LM33	No change in complete blood count LM33_P1 reduced lymphocyte count in the blood 536_P1 elevated IFN-γ, IL-12, MCP-1 and keratinocyte chemoattractant in the lung at 17 hours post-administration in lungs but not in the blood	[43]
Pf4	*P. aeruginosa* PAO1	Intratracheal/ C57BL/6 mice	1.5×10^7 PFU	Reduced CXCL1, IL-17, IL-1β, IL-17 and IL-10 in the bronchoalveolar lavage fluid IL-12p40 in the bronchoalveolar lavage fluid was significantly higher at 48 hours post-administration compared to PAO1-infection alone No significant differences in the number of macrophages after phage treatment	[44]

(continued)

Table 1 (continued)

Phage	Pathogen	Route of administration/ animal	Dose	Immunological effects	Reference
Phage cocktail (contains: PBAB08, PBAB25, PBAB68, PBAB93)	*A. baumanii* (14 clinical strains)	Intraperitoneal, intranasal, oral/ BALB/c mice	10^9 PFU	Minimal increase in serum cytokine (IL-1a, IL-1b, IL-2, IL-4, IL-6, IL-10, IL-12, IL-17A, TNF-a, G-CSF) in all routes. Increased GM-CSF in all routes of administration	[8]
PAK_P1	*P. aeruginosa* PAKlumi	Intranasal/ BALB/c, C57B/6/J, Rag2$^{-/-}$Il2rg$^{-/-}$, Myd88$^{-/-}$ mice	10^8 PFU	No significant increase in cytokine production in lung tissue except IFN-y and TNF-α Synergy observed between the phage and the host's immune system	[38]

difference between phage therapy and standard-of-care treatment due to the instability of the phages, which led to the administration of a subtherapeutic dose [26]. This highlights another hurdle that phage therapy must overcome to be actively used clinically. Recent research suggests that adding a stabiliser to the formulation can minimise the loss of viability over long-term storage [45–47]. Naturally, the safety of these excipients must be investigated before clinical use.

8.5.1 Liquid Aerosols

The significance of the PhagoBurn trial rested on the fact that it was the first randomised controlled trial investigating a lytic phage cocktail that was produced according to good manufacturing practice and was approved by national health regulators in Belgium, France, and Switzerland. The shortfall of the trial was due to the unexpected decline by 10^4–10^5 PFU/mL in titre post-manufacturing, which resulted in a much lower than the expected-administered dose of 10^6 PFU/mL [26]. Post-trial stability study demonstrated that the initial titre reduction occurred 15 days after manufacturing. A major conclusion of the trial was that decreasing the number of phage types and improving the stability were needed for future trials. As the therapy was intended for IV injection, the cocktail was a liquid formulation that was thought to be stable under refrigeration [48, 49]. However, Duyvejonck et al. [46] recently confirmed that at 4 °C phages with a high titre of 10^9 PFU/mL were relatively more stable than the 10^7 PFU/mL stock. This demonstrated that phage titre decreases over time even when stored refrigerated, especially at lower starting titres [50]. Although ancillary stability testing prior to the PhagoBurn trial found that the individual phages of the cocktail were very stable at a titre of ≥10^9 PFU/mL for over 24 months, the authors reported that there was a 1000-fold reduction of the stock within 15 days of manufacture [26]. The starting overall titre of the cocktail was 10^9 PFU/mL, meaning that the average titre for each of the 12 phages was 8.3×10^7 PFU/mL, which was lower than that tested for their individual stability. As discussed above, phage OMKO1 had synergy with ceftazidime in treating MDR *P. aeruginosa* infection [34]. Blazanin et al. [51] tested the effect of various conditions, including elevated temperature, varying saline and urea concentrations, that phages may encounter in vitro and in vivo to study their stability. Interestingly, although moderate titre reduction was only observed from 65 to 75 °C, the plaque morphology of all heat-treated phages changed. Further investigation demonstrated that the growth ability of phages was significantly decreased compared to the control regardless of the duration of thermal exposure at 70 °C [51]. A change in plaque morphology may not necessarily equate to a change in the phenotype of the phage, as transmission electronic microscopy (TEM) of phages (TS3, TS6, TS13) treated at 70 °C for 30 minutes showed no morphological difference compared to the control [52]. On the other hand, phage OMKO1 may not be a thermally stable phage as phages TS3, TS6 and TS13 were specifically isolated for their thermal-resistant properties [52]. Furthermore, urea stress tolerance over time revealed that the survivability of phage OMKO1 was unaffected at lower urea concentrations of 1–2 M and was not significantly different to that without urea [51]. However, phage survival reduced quickly over the 90 minutes in 3–4 M urea [51]. Unlike heat and high urea concentrations, high saline concentrations did not impact phage infectivity [51]. Phages Acibel004, PNM, 14/1, and ISP stored in normal saline at 4 °C at 10^9 PFU/mL for over 300 days showed no change in phage infectivity [46]. Since saline is commonly used in pharmaceutical products, stability in saline makes it a viable vehicle for formulating therapeutic phages.

Liquid droplets for lung delivery can be produced by jet nebulisation, vibrating mesh nebulisation, and colliding jets (Respimat® inhaler) [53], which have been applied to aerosolising phages. Comparison between these three methods showed differences in titre loss post-nebulisation [54]. Jet nebulised active PEV44 phages was significantly lower at 5.4×10^4 PFU/minute, compared to those aerosolised from a vibrating-mesh nebuliser (3.3×10^8 PFU/minute) and soft-mist inhaler (4.6×10^6 PFU/minute) [54]. The possible reason for such a discrepancy in titre loss between the different nebulisers is likely due to the intense shearing in jet nebulisers that damaged the phages [53, 54]. Infectivity correlates with the structural integrity of phages as the extent of capsid and tail detachment was associated with the titre loss of PEV44 [55]. The degree of reduction was higher in *Siphoviridae* phages (e.g. D29) and *Myoviridae* (e.g. PEV44) [55] due to their longer tails than those of *Podoviridae* phages [56]. The examination of structural differences in the infectivity of nebulised phages revealed that the long-tailed D29 and PEV40 suffered from more head-tail detachment that correlated with the reduction in the titre [56]. On the other hand, jet nebulisation had minimal impact on PEV2 as its short tail is less likely to break. Therefore, structural differences between phages and the choice of the nebuliser are important factors to consider for respiratory delivery.

FluMist® is the first intranasal vaccine against seasonal influenza. It contains the live attenuated influenza virus and is stable under refrigeration [57]. The reason for its stability is due to having been frozen at a temperature below the glass transition temperature [58]. Although phages are also viruses and may also be formulated similarly, cold-chain maintenance is difficult and costly. Therefore, it is worthwhile to investigate formulations that are stable at room temperature, such as solid dosage forms.

8.5.2 Dry Powder Aerosols

Dry powder formulations for inhalation are popular owing to their higher chemical stability than liquids and storage at ambient temperature [59]. Lyophilisation has an immediate but minor negative impact on the infectivity of specific phages, depending on the excipients used [45, 60]. Storage of the lyophilised formulation for another 27 months only showed an insignificant decrease in the titre [60]. Storage for 8 years at 4 °C did not reduce the titre significantly ($n = 1$), except when 0.8 M trehalose was used as an excipient, for which a 1.75-log titre reduction was observed [46]. However, lyophilised powders have ill-defined particle shapes and sizes so they may not be suitable for inhalation.

Spray drying is a one-step process that can produce inhalable phage powders with only a moderate titre loss [47, 61, 62]. The spray drying temperature may negatively impact phage stability, in turn affecting phage infectivity and aerosol performance. A low drying temperature is required to minimise this impact [63, 64]. The storage temperature may also influence phage viability [45, 62]. Storage of spray-dried phages PEV2 and PEV40 at 4 and 20 °C in vacuum packaging yielded different results that were formulation-dependent [62]. For PEV2, formulations containing 70% trehalose and 30% leucine were stable with no further titre reduction up to 1 year at both 4 and 20 °C. On the other hand, those containing 60% trehalose and 40% leucine showed temperature-dependent titre reduction [62]. Similar to the lyophilisation of phage ISP [60], spray drying PEV2 and PEV40 resulted in an immediate reduction of 0.7 and 0.2 log PFU/mL, respectively, regardless of the different excipient combinations [62]. PEV2 titres were not further affected when stored at 4 and 20 °C if the formulation contained 70% trehalose and 30% leucine [62]. However, when the trehalose content decreased to 60%, PEV2 was stable when stored at 4 °C for up to 12 months but its viability decreased significantly when stored at 20 °C after 6 months. PEV40 displayed more variability in the titre and for all formulations and temperatures studied but its titre was stable for up to 3 months with no further degradation [62]. Currently, lactose is approved as an excipient for inhalation. On the other hand, trehalose and leucine are yet to be approved for inhalation, although they are Generally Recognised as Safe excipients for other types of formulations. They have been used in experimental inhaled powders as they can improve the physical stability of the drug/powder and have relatively low toxicity [65]. Flexible small sugars such as trehalose protects against heat and water-removal which are desirable traits for phage formulations as the proteins require hydrogen bonds for structural stability [66, 67] by decreasing molecular mobility [68]. Spray-dried phages PEV20 and PEV61 showed different storage characteristics depending on the formulation, which consisted of lactose (55–90%) and leucine (45–10%) [47]. In vitro cell studies demonstrated that the combination of phage, lactose and leucine did not negatively affect the viability of human epithelial cells and macrophages, supporting the safety of these excipients [47]. A high lactose content of 80% and above increased the stability of the phage powders at ambient temperature when they were packed in an aluminium pouch at 15% relative humidity (RH) and stored at 20 °C/60% RH [47]. This resulted in minimal titre loss over 12 months for phages PEV1, PEV61 and PEV20. The extent of titre reduction and the duration of stability during storage varied between phage species. PEV20 was more prone to inactivation compared to PEV1 and PEV61. A low leucine content (10%) was shown to be sufficient to protect phage powders from moisture-induced degradation. Another major factor to consider is the difference between the glass transition temperature (T_g) and storage temperature (T_s), i.e. $T_g - T_s$ [68]. Higher storage RH lowers the T_g, consequently lowering the $T_g - T_s$ value, resulting in phage titre reduction regardless of the storage temperature. Chang et al. [68] demonstrated that the powder adsorbed more moisture at 33% RH than at 15% RH. Since water is a strong plasticiser, protein mobility was increased and consequently phage stability decreased. The spray-dried phages remained stable when $T_g - T_s$ was >46 °C, hence it is vital to consider T_g when selecting the storage conditions for phage powders. To facilitate storage, phage powders should be stable at room temperature (20–25 °C), meaning that T_g should be at least 71 °C, which may be achievable by choosing excipient/s with a high T_g and/or keeping the storage RH low.

There is currently a lack of longer-term stability studies lasting 12 months or longer, given that the shelf-life is very important if phage therapy is to be extensively utilised in the future. The formulation of a personalised cocktail from a phagogram in clinical settings will only be possible if there are viable phage products readily available.

8.6 Future Directions

Currently, there is no regulatory framework governing the manufacture and use of phage products but Belgium has established a phage monograph for magistral preparations endorsed by their Federal Agency for Medicines and Health Products [69]. However, this is purposed for patient-specific compassionate use rather than scaled-up therapeutics such as antibiotics. Phagoburn was the first clinical trial of a GMP-grade phage cocktail and its protocols were approved by the French Food Safety Agency [26]. Similar to biologics, phage manufacturers must be GMP compliant and require regulations to ensure the safety, purity, and efficacy of the product. It is very likely that the costs involved in the production of phage products will exceed that of antibiotics, which would incur a financial strain on the stakeholders. However, given the increasing incidence of resistant infections, phage therapy could be a potentially life-saving option, especially when used concomitantly with antibiotics.

High-quality clinical trials of phage therapy are required. Due to the lack of pharmacokinetic and pharmacodynamic studies, there is limited understanding of the optimal dose of phages [6, 14]. Their dosage may be species-specific but the common dose tested in published studies is 10^7–10^9 PFU/mL [6, 8, 9, 12, 20, 34, 70]. An in vivo study in neutropenic BALB/c mice using an intratracheal anti-pseudomonas phage, PEV31, demonstrated dose-dependent responses in phage replication in situ, the number of emergent phage-resistant bacteria, and the level of proinflammatory cytokines (IL-1β, IL-6, TNF-α) [71]. However, the extent of bacterial load clearance was independent of the dose of phage delivered [71]. It is unclear whether this finding is phage-dependent or whether it is applicable to all phages because there is insufficient data on dosing for all routes of phage administration [14]. Nevertheless, there is an increasing number of registered clinical trials on phages [72] that can mend the current gaps in knowledge. A major safety concern of administering phages is their purity as endotoxins released from the host bacteria as part of the manufacturing process will negatively impact the patient. Furthermore, bacteriolysis by phages during treatment may rapidly release endotoxins which may lead to a cascade of pro-inflammatory responses. Although adverse events are seldom reported in the literature and phage therapy is generally considered to be safe [14], a systematic review reports that the quality of evidence on the safety of phage therapy is low or moderate because most published studies were case studies of individual patients [73]. FDA limits the endotoxin levels for the single parenteral dose to 5 endotoxin units (EU)/kg of body weight per hour of administration. Following current GMP is the safest method in producing phages for clinical use, as confirmed by Fabijan et al. [70], who reported no safety concerns in treating severe *S. aureus* infections.

Re-emergence of phage therapy for MDR infections has highlighted the knowledge gaps that are preventing its wide clinical use. High-quality clinical trials are undoubtedly needed to investigate their safety, quality, and efficacy.

Acknowledgement The authors received no financial support for the research, authorship, and publication of this article.

References

1. Keen E. A century of phage research: bacteriophages and the shaping of modern biology. Bioessays. 2015;37(1):6–9.
2. Summers W. Félix Hubert d'Herelle (1873–1949): history of a scientific mind. Bacteriophage. 2016;6(4):e1270090.

3. O'Neill J. Antimicrobial resistance: tackling a crisis for the health and wealth of nations. London: Review on Antimicrobial Resistance; 2014.
4. Principi N, Silvestri E, Esposito S. Advantages and limitations of bacteriophages for the treatment of bacterial infections. Front Pharmacol. 2019;10:513.
5. Mu A, McDonald D, Jarmusch A, Martino C, Brennan C, Bryant M, et al. Assessment of the microbiome during bacteriophage therapy in combination with systemic antibiotics to treat a case of staphylococcal device infection. Microbiome. 2021;9:92.
6. Guillon A, Pardessus J, L'Hostis G, Fevre C, Barc C, Dalloneau E, et al. Inhaled bacteriophage therapy in a porcine model of pneumonia caused by *Pseudomonas aeruginosa* during mechanical ventilation. Br J Pharmacol. 2021;178:3829–42.
7. Brives C, Pourraz J. Phage therapy as a potential solution in the fight against AMR: obstacles and possible futures. Palgrave Commun. 2020;6:100.
8. Cha K, Oh H, Jang J, Jo Y, Kim W, Ha G, et al. Characterisation of two novel bacteriphages infecting multidrug-resistant (MDR) *Acinetobacter baumannii* and evaluation of their therapeutic efficacy in vivo. Front Microbiol. 2018;9:696.
9. Chadha P, Katare O, Chhibber S. In vivo efficacy of single phage versus phage cocktail in resolving burn wound infection in BALB/c mice. Microb Pathog. 2016;99:68–77.
10. Jeon J, Park J, Yong D. Efficacy of bacteriophage treatment against carbapenem-resistant *Acinetobacter baumannii* in *Galleria mellonella* larvae and a mouse model of acute pneumonia. BMC Microbiol. 2019;19(1):70.
11. Waters E, Neill D, Kaman B, Sahota J, Mrj C, Winstanley C, et al. Phage therapy is highly effective against chronic lung infections with *Pseudomonas aeruginosa*. Thorax. 2017;72:666–7.
12. Aslam S, Lampley E, Wooten D, Karris M, Benson C, Strathdee S, et al. Lessons learned from the first 10 consecutive cases of intravenous bacteriophage therapy to treat multidrug-resistant bacterial infections at a single center in the United States. Open Forum Infect Dis. 2020;7(9):ofaa389.
13. Law N, Logan C, Yung G, Langlais Furr C, Lehman S, Morales S, et al. Successful adjunctive use of bacteriophage therapy for treatment of multidrug-resistant *Pseudomonas aeruginosa* infection in a cystic fibrosis patient. Infection. 2019;47:665–8.
14. Suh G, Lodise T, Tamma P, Knisely J, Alexander J, Aslam S, et al. Considerations for the use of phage therapy in clinical practice. Antimicrob Agents Chemother. 2022;66(3):e02071-21.
15. Nale J, Clokie M. Preclinical data and safety assessment of phage therapy in humans. Curr Opin Biotechnol. 2021;68:310–7.
16. Prazak J, Valente L, Iten M, Federer L, Grandirard D, Soto S, et al. Benefits of aerosolised phages for the treatment of pneumonia due to methicillin-resistant *Staphylococcus aureus*: an experimental study in rats. J Infect Dis. 2022;225(8):1452–9.
17. Prazak J, Valente L, Iten M, Grandirard D, Leib S, Jakob S, et al. Nebulised bacteriophages for prophylaxis of expiermental ventilator-associated penumonia due to methicillin-resistant *Staphylococcus aureus*. Crit Care Med. 2020;48(7):1042–6.
18. Herridge W, Shibu P, O'Shea J, Brook T, Hoyles L. Bacteriophages of *Klebsiella* spp., their diversity and potential therapeutic uses. J Med Microbiol. 2020;69(2):176–94.
19. Gan L, Fu H, Tian Z, Cui J, Yan C, Xue G, et al. Bacteriophage effectively rescues pneumonia caused by prevalent multidrug-resistant *Klebsiella pneumoniae* in the early stage. Microbiol Spectr. 2022;10(5):e0235822.
20. Cao F, Wang X, Wang L, Li Z, Che J, Wang L, et al. Evaluation of the efficacy of a bacteriophage in the treatment of pneumonia induced by multidrug resistance *Klebsiella pneumoniae* in mice. Biomed Res Int. 2015;2015:752930.
21. Anand T, Virmani N, Kumar S, Mohanty A, Pavulray S, Bera B, et al. Phage therapy for treament of virulent *Klebsiella penumoniae* infection in a mouse model. J Glob Infect Dis. 2020;21:34–41.
22. Guillon A, Pardessus J, L'Hostis G, Fevre C, Barc C, Dalloneau E, et al. Inhaled bacteriophage therapy in a procine model of pneumonia caused by *Pseudomonas aeruginosa* during mechanical ventilation. Br J Pharmacol. 2021;178:3829–42.

23. ClinicalTrials.gov. Study to evaluate the safety, phage kinetics, and efficacy of inhaled AP-PA02 in subjects with non-cystic fibrosis bronchiectasis and chronic pulmonary pseudomonas aeruginosa nfection (Tailwind). Armata Pharmaceuticals, Inc.; 2022 [updated 24 May 2023].
24. ClinicalTrials.gov. Ph 1/2 study evaluating safety and tolerability of inhaled AP-PA02 in subjects with chronic pseudomonas aeruginosa lung infections and cystic fibrosis (SWARM-Pa). Armata Pharmaceuticals, Inc.; 2020 [updated 22 Dec 2022]. Available from: https://clinicaltrials.gov/ct2/show/NCT04596319?term=phage+inhale&cond=Pseudomonas+Aeruginosa+Infection&draw=2&rank=3.
25. ClinicalTrials.gov. Nebulized bacteriophage therapy in cystic fibrosis patients with chronic pseudomonas aeruginosa pulmonary infection. BiomX, Inc.; 2021 [updated 18 Apr 2023]. Available from: https://clinicaltrials.gov/ct2/show/NCT05010577?term=phage+inhale&cond=Pseudomonas+Aeruginosa+Infection&draw=2&rank=2.
26. Jault P, Leclerc T, Jennes S, Pirnay J, Que Y, Resch G, et al. Efficacy and tolerability of a cocktail of bacteriophages to treat burn wounds infected by *Pseudomonas aeruginosa* (Phagoburn): a randomised, controlled, double-blind phase 1/2 trial. Lancet Infect Dis. 2019;19:35–45.
27. Zeynali Kelishomi F, Khanjani S, Fardsanei F, Sarabi H, Nikkhahi F, Dehghani B. Bacteriophages of *Mycobacterium tuberculosis*, their diversity, and potential therapeutic uses: a review. BMC Infect Dis. 2022;22:957.
28. Carrigy N, Larsen S, Reese V, Pecor T, Harrison M, Kuehl P, et al. Prophylaxis of *Mycobacterium tuberculosis* H37Rv infection in a preclinical mouse model via inhalation of nebulised bacteriophage D29. Antimicrob Agents Chemother. 2019;63:12.
29. Dedrick R, Smith B, Cristinziano M, Freeman K, Jacobs-Sera D, Belessis Y, et al. Phage therapy of *Mycobacterium* infections: compassionate use of phages in 20 patients with drug-resistant mycobacterial diseases. Clin Infect Dis. 2023;76(1):103–12.
30. Oechslin F, Piccardi P, Mancini S, Gabard J, Moreillon P, Entenza J, et al. Synergistic interaction between phage therapy and antibiotics clears *Pseudomonas aeruginosa* infection in endocarditis and reduces virulence. J Infect Dis. 2017;215(5):703–12.
31. Lin Y, Chang R, Britton W, Morales S, Kutter E, Chan H. Synergy of nebulised phage PEV20 and ciprofloxacin combination against *Pseudomonas aeruginosa*. Int J Pharm. 2018;551(1–2):158–65.
32. Lin Y, Quan D, Chang R, Chow M, Wang Y, Li M, et al. Synergistic activity of phage PEV20-ciprofloxacin combination powder formulation – a proof-of-prinicple study in a *P. aeruginosa* lung infection model. Eur J Pharm Biopharm. 2021;158:166–71.
33. Gurney J, Pradier L, Griffin J, Gougat-Barbera C, Chan B, Turner P, et al. Phage steering of antibiotic-resistance evolution in the bacterial pathogen *Pseudomonas aeruginosa*. Evol Med Public Health. 2020;2020(1):148–57.
34. Chan B, Turner P, Kim S, Mojibian H, Elefteriades J, Narayan D. Phage treatment of an aortic graft infected with *Pseudomonas aeruginosa*. Evol Med Public Health. 2018;2018(1):60–6.
35. Chan B, Sistrom M, Wertz J, Kortright K, Narayan D, Turner P. Phage selection restores antibiotic sensitivity in MDR *Pseudomonas aeruginosa*. Sci Rep. 2016;6:26717.
36. Liu C, Hong Q, Chang R, Kwok P, Chan H. Phage-antibiotic therapy as a promising strategy to combat multidrug-resistant infections and to enhance antimicrobial efficiency. Antibiotics. 2022;25(11):570.
37. Yang Y, Shen W, Zhong Q, Chen Q, He X, Baker J, et al. Development of a bacteriophage cocktail to constrain the emergence of phage-resistnat *Pseudomonas aeruginosa*. Front Microbiol. 2020;11:327.
38. Roach D, Leung C, Henry M, Morello E, Singh D, Di Santo J, et al. Synergy between the host immune system and bacteriophage is essential for successful phage therapy against an acute respiratory pathogen. Cell Host Microbe. 2017;22:38–47.
39. Gordillo Altamirano F, Barr J. Unlocking the next generation of phage therapy: the key is in the receptors. Curr Opin Biotechnol. 2021;68:115–23.

40. Tisoncik J, Korth M, Simmons C, Farrar J, Martin T, Katze M. Into the eye of the cytokine storm. Microbiol Mol Biol Rev. 2012;76(1):16–32.
41. Suntharalingam G, Perry M, Ward S, Brett S, Castello-Cortes A, Brunner M, et al. Cytokine storm in a phase 1 trial of the anti-CD28 monoclonal antibody TGN1412. N Eng J Med. 2006;355:1018–28.
42. Pabary R, Singh C, Morales S, Bush A, Alshafi K, Bilton D, et al. Antipseudomonal bacteriophage reduces infective burden and inflammatory response in murine lung. Antimicrob Agents Chemother. 2015;60(2):744–51.
43. Dufour N, Delattre R, Chevallereau A, Richard J, Debarbieux L. Phage therapy of pneumonia is not associated with an overstimulation of the inflammatory response compared to antibiotic treatment in mice. Antimicrob Agents Chemother. 2019;63(8):e00379-19.
44. Secor P, Michaels L, Smigiel K, Rohani M, Jennings L, Hisert K, et al. Filamentous bacteriophage produced by *Pseudomonas aeruginosa* alters the inflammatory reponse and promotes noninvasive infection in vivo. Infect Immun. 2017;85:e00648-16.
45. Gonzalez-Menendez E, Fernandez L, Gutierrez D, Rodriguez A, Martinez B, Garcia P. Comparative analysis of different preservation techniques for the storage of *Staphylococcus* phages aimed for the industrial development of phage-based antimicrobial products. PLoS One. 2018;13(10):e0205728.
46. Duyvejonck H, Merabishvili M, Vaneechoutte M, de Soir S, Wright R, Friman V, et al. Evaluation of the stability of bacteriophages in different solutions suitable for the production of magistral preparations in Belgium. Viruses. 2021;13(5):865.
47. Chang R, Wallin M, Kutter E, Morales S, Britton W, Li J, et al. Storage stability of inhalable phage powders containing lactose at ambient conditions. Int J Pharm. 2019;560:11–8.
48. Jonczyk E, Klak M, Miedzybrodzki R, Gorski A. The influence of external factors on bacteriophages – review. Folia Microbiol. 2011;56:191–200.
49. Chang R, Morales S, Okamoto Y, Chan H. Topical application of bacteriophages for treatment of wound infections. Transl Res. 2020;220:153–66.
50. Golec P, Dabrowski K, Hejnowicz M, Gozdek A, Los J, Wegrzyn G, et al. A reliable method for storage of tailed phages. J Microbiol Methods. 2011;84(3):486–9.
51. Blazanin M, Lam W, Vasen E, Chan B, Turner P. Decay and damage of therapeutic phage OMKO1 by envrionmental stressors. PLoS One. 2022;17(2):e0263887.
52. Park H, Kim J, Kim M, Park Y, Ryu S. Development of new strategy combining heat treatment and phage cocktail for post-contamination prevention. Food Res Int. 2021;145:110415.
53. Martin A, Finlay W. Nebulisers for drug delivery to the lungs. Expert Opin Drug Deliv. 2014;12(6):889–900.
54. Carrigy N, Chang R, Leung S, Harrison M, Petrova Z, Pope W, et al. Anti-tuberculosis bacteriophage D29 delivery with a vibrating mesh nebuliser, jet nebuliser, and soft mist inhaler. Pharm Res. 2017;34(10):2084–96.
55. Astudillo A, Leung S, Kutter E, Morales S, Chan H. Nebulisation effects on structural stability of bacteriophage PEV 44. Eur J Pharm Biopharm. 2018;125:124–30.
56. Leung S, Carrigy N, Vehring R, Finlay W, Morales S, Carter E, et al. Jet nebulisation of bacteriophages with different tail morphologies – structural effects. Int J Pharm. 2019;554:322–6.
57. AstraZeneca. Storage and handling of FLUMIST QUADRIVALENT. 2022 [updated Jan 2023]. Available from: https://www.flumistquadrivalenthcp.com/storage-handling.html.
58. Tlaxca J, Ellis S, Remmele R Jr. Live attenuated and inactivated viral vaccine formulation and nasal delivery: potential and challenges. Adv Drug Deliv Rev. 2015;93(1):56–78.
59. Shetty N, Cipolla D, Park H, Zhou Q. Physical stability of dry powder inhaler formulations. Expert Opin Drug Deliv. 2020;17(1):77–96.
60. Merabishvili M, Vervaet C, Pirnay J, De Vos D, Verbeken G, Mast J, et al. Stability of *Staphylococcus aureus* phage ISP after freeze-drying (lyophilisation). PLoS One. 2013;8(7):e68797.
61. Chang R, Wong J, Mathai A, Morales S, Kutter E, Britton W, et al. Production of highly stable spray dried phage formulations for treatment of *Pseudomonas aeruginosa* lung infection. Eur J Pharm Biopharm. 2017;121:1–13.

62. Leung S, Parumasivam T, Nguyen A, Gengenbach T, Carter E, Carrigy N, et al. Effect of storage temperature on the stability of spray dried bacteriophage powders. Eur J Pharm Biopharm. 2018;127:213–22.
63. Chang R, Wallin M, Lin Y, Leung S, Wang H, Morales S, et al. Phage therapy for respiratory infections. Adv Drug Deliv Rev. 2018;133:76–86.
64. Lin Y, Chang R, Britton W, Morales S, Kutter E, Li J, et al. Storage stability of phage-ciprofloxacin combination powders against *Pseudomonas aeruginosa* respiratory infections. Int J Pharm. 2020;591:119952.
65. Zillen D, Beugeling M, Hinrichs W, Frijlink H, Grasmeijer F. Natural and bioinspired excipients for dry powder inhalation formulations. Curr Opin Colloid Interface Sci. 2021;56:101497.
66. Vinner G, Rezaie-Yazdi Z, Leppanen M, Stapley A, Leaper M, Malik D. Microencapsulation of *Salmonella*-specific bacteriophage Felix O1 using spray-drying in a pH-responsive formulation and direct compression tableting of powders into a solid oral dosage form. Pharmaceuticals. 2019;12:43.
67. Vandenheuvel D, Meeus J, Lavigne R, Van den Mooter G. Instability of bacteriophages in spray-dried trehalose powders is caused by crystallisation of the matrix. Int J Pharm. 2014;472(1–2):202–5.
68. Chang R, Kwok P, Khanal D, Morales S, Kutter E, Li J, et al. Inhalable bacteriophage powders: glass transition temperature and bioactivity stabilisation. Bioeng Transl Med. 2020;5(2):e10159.
69. Bretaudeau L, Tremblais K, Aubrit F, Meichenin M, Arnaud I. Good manufacturing practice (GMP) compliance for phage therapy medicinal products. Front Microbiol. 2020;11:1161.
70. Fabijan A, Lin R, Ho J, Maddocks S, Zakour N, Iredell R, et al. Safety of bacteriophage therapy in severe *Staphylococcus aureus* infection. Nat Microbiol. 2020;5:465–72.
71. Chang R, Chow M, Wang Y, Liu C, Hong Q, Morales S, et al. The effects of different doses of inhaled bacteriophage therapy for *Pseudomonas aeruginosa* pulmonary infections in mice. Clin Microbiol Infect. 2022;28(7):983–9.
72. Furfaro L, Payne M, Chang B. Bacteriophage therapy: clinical trials and regulatory hurdles. Front Cell Infect Microbiol. 2018;8:376.
73. Uyttebroek S, Chen B, Onsea J, Ruythooren F, Debaveye Y, Devolder D, et al. Safety and efficacy of phage hterapy in difficult-to-treat infections: a systematic review. Lancet Infect Dis. 2022;22(8):E208–20.

Pharmacokinetics of Inhaled Medications – What Do We Know About Biological Macromolecules?

Audrey Huili Lim, Jinxin Zhao, Nusaibah Abdul Rahim, Jing Zhao, Haiting Liu, Xiaoyan Yang, and Yu-Wei Lin

Abstract Over the last decades, biological macromolecules have attracted growing attention and have become an important therapeutic modality. Biological macromolecules are peptide- and protein-based therapeutics with unique pharmacokinetic (PK) and pharmacodynamic (PD) properties. Due to its poor bioavailability, biological macromolecules are usually administered via the parenteral route. Pulmonary administration of biological macromolecules is increasingly employed for treating various diseases, including respiratory viral infections. Due to its complex PK/PD, clinical development of inhaled biological macromolecules remains challenging.

A. H. Lim (✉) · N. A. Rahim
Malaya Translational and Clinical Pharmacometrics Group, Faculty of Pharmacy, University of Malaya, Kuala Lumpur, Malaysia

Department of Clinical Pharmacy and Pharmacy Practice, Faculty of Pharmacy, University of Malaya, Kuala Lumpur, Malaysia

J. Zhao
Malaya Translational and Clinical Pharmacometrics Group, Faculty of Pharmacy, University of Malaya, Kuala Lumpur, Malaysia

Infection Program and Department of Microbiology, Monash Biomedicine Discovery Institute, Monash University, Melbourne, VIC, Australia

J. Zhao · H. Liu · X. Yang
Department of Neonatology, West China Second University Hospital of Sichuan University, Chengdu, Sichuan, People's Republic of China

Y.-W. Lin (✉)
Malaya Translational and Clinical Pharmacometrics Group, Faculty of Pharmacy, University of Malaya, Kuala Lumpur, Malaysia

Department of Clinical Pharmacy and Pharmacy Practice, Faculty of Pharmacy, University of Malaya, Kuala Lumpur, Malaysia

Infection Program and Department of Microbiology, Monash Biomedicine Discovery Institute, Monash University, Melbourne, VIC, Australia
e-mail: yuwei.lin@um.edu.my

J. K. W. Lam, P. C. L. Kwok (eds.), *Respiratory Delivery of Biologics, Nucleic Acids, and Vaccines*, AAPS Introductions in the Pharmaceutical Sciences 8,
https://doi.org/10.1007/978-3-031-47567-2_9

In this context, pharmacometrics using quantitative modeling and simulation techniques offer critical insight in understanding the complex PK/PD and providing guidance at various stages throughout the development of inhaled biological macromolecules. This chapter provides an overview of the key pulmonary PK processes and the available modeling and simulation (M&S) tools that can be used to assess and understand PK in the clinical development of inhaled medications, including biological macromolecules.

Keywords Biologics · Mathematical model · Model-informed drug development · Pharmacokinetics · Pharmacometrics

1 Introduction

Respiratory diseases represent a major global health burden, with approximately 70,000 people living with cystic fibrosis, 250 million with chronic obstructive pulmonary disease (COPD), and 20 million with asthma [1–3]. Furthermore, the annual global death toll from lung cancers and/or pneumonia represents the single largest death, accounting for more than 1.8 million deaths annually [4]. This figure is expected to rise sharply due to the recent outbreak of the COVID-19 pandemic. The respiratory drug market is growing at a rate of 4–6% annually; many new therapeutics, including biologics, are currently under development [5].

Biologics refers to a diverse group of large and complex molecules manufactured in, extracted from, or semi-synthesized from biological sources [6]. Generally, biopharmaceuticals range from vaccines and growth factors to enzymes, fusion proteins, hormones, and monoclonal antibodies. Biologics have high commercial potential as they are often associated with high specificity and high target binding affinity. Nonetheless, biologics are also associated with several limitations, for example, fragility, risk of immune response, high cost, and often require specific storage conditions to prevent degradation [6–9]. Biological macromolecules including therapeutic peptides, proteins, and antibodies, represent the largest group of biological drugs and are the focus of this chapter.

Successful delivery of biological macromolecules via a non-invasive route remains elusive due to their high polarity and large molecular size [6]. Various non-invasive routes of administration have been investigated for the delivery of biological macromolecules such as oral, nasal, inhalation, and transdermal routes. Pulmonary administration is an attractive route because of the large surface area, extensive vascularization, and high tissue permeability, which enable rapid absorption (depending on the modality) and the potential for rapid onset of activity if the target is local. Moreover, the inhaled route is a comparatively easier form of administration and allows potential self-administration by patients [10]. Pulmonary delivery also offers the advantage of achieving high concentrations in the lungs. To achieve the same drug concentrations in the lungs after systemic drug administration, higher doses would normally be needed compared with inhalation, which would be associated with an increased risk of side effects [11, 12]. As such, aerosolized biological macromolecules are preferable for the treatment of lung diseases such as asthma, lung infection, and COPD [6]. In addition, there is little first-pass metabolism and proteolytic enzyme activity in the lungs [11, 13].

Clinical pharmacology is the study of the interactions between drugs and the human body and comprises two broad divisions: pharmacokinetics (PK) and pharmacodynamics (PD) [14]. PK is defined as the study of the time course of drug absorption, distribution, metabolism, and elimination (i.e., ADME) while PD is defined as the relationship between drug concentration and the resulting therapeutic effect [15]. Simplistically, PK is the study of what the body does to the drug, whereas PD is the study of what the drug does to the body. Characterization of the relationship between PK and PD is imperative in the discovery and development of new molecules in the pharmaceutical industry. Insights from PK and PD analyses are used by researchers to design better clinical studies, and by clinicians to advise treatment options [16, 17]. The role of mathematical modeling to assess and understand the deposition, PK, and PD of inhaled drugs has gained significant interest over the last decade [18]. This chapter provides an overview of the key pulmonary PK processes and the available modeling and simulation (M&S) tools that can be used to assess and understand PK in the clinical development of inhaled medication, including biological macromolecules.

2 Key Determinants of Pulmonary PK

To understand PK following inhalation, a thorough understanding of the key kinetic processes is crucial (Table 1 and Fig. 1). It should be noted that the key factors governing the PK of inhaled biological macromolecules are currently not fully understood. Our current understanding of the PK of inhaled biological molecules is very limited and available literature is mostly based on nonclinical models with very limited clinical utility. This section provides an overview of the key factors impacting pulmonary PK processes based on our understanding of small molecules. It is expected that these factors are also applicable to inhaled biological macromolecules.

Table 1 Key determinants that affect pulmonary and systemic PK following inhalation

Kinetic process	Key determinants
Pulmonary deposition	Lung morphology, inhalation techniques, formulation, aerosol velocity, geometry, particle size and particle density
Pulmonary dissolution	Physiochemical property of the drug/formulation (e.g., solubility) and the physiology of the lungs
Absorption	Bronchial mucus layer, alveolar surfactant, lung epithelium, lipophilicity, the degree of ionization, molecular weight, drug transporters, areas of the lungs
Distribution	Degree of ionization, lipophilicity
Metabolism	Expression levels and patterns of the metabolic enzymes, lung macrophages
Elimination	Particle size, solubility, deposition site

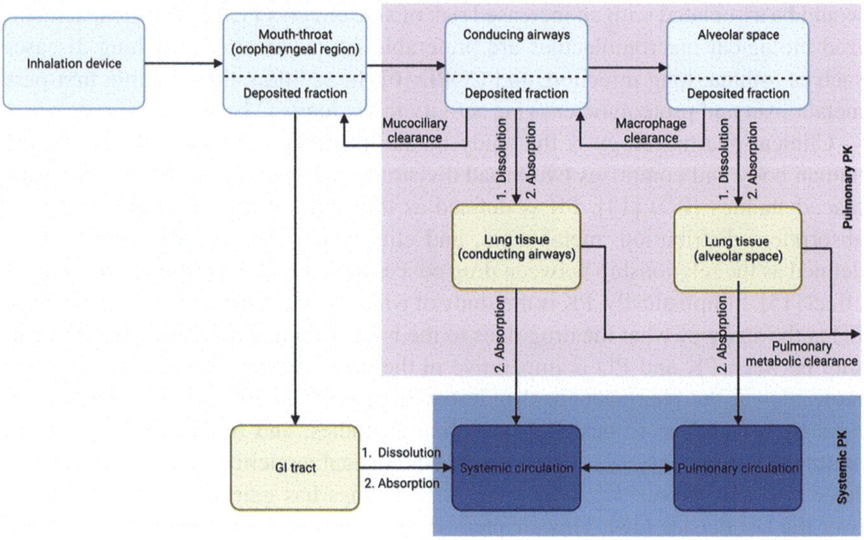

Fig. 1 Schematic representation of the kinetic processes following inhalation. (Adapted from Borghart et al. [18]). Light purple boxes = particle/droplet deposition compartments; light orange boxes = absorption compartments; purple boxes = blood/plasma compartments; purple area = processes associated with pulmonary; blue area = processes associated with the systemic PK

2.1 Pulmonary Physiology

The lungs are a pair of complex, heterogeneous, pyramid-shaped organs situated in the thoracic cavity (Fig. 2). They are the foundational organs of the respiratory system and consist of two main functional areas: gas exchange parenchyma (e.g., alveolar sacs) and conducting airways (e.g., bronchioles) [19–21].

Conducting airways bifurcate approximately 17 times, getting smaller with each sequential bifurcation [23–25]. The airways are lined with pseudo-stratified columnar epithelia, which are mostly composed of basal cells, secretory cells (e.g., goblet and serous cells), and ciliated cells [24, 25]. The walls of conducting airways are too thick for diffusion, and this region does not contain alveoli to facilitate oxygen-carbon dioxide exchange. The main functions of the conducting airways are to transmit air to the lung, as well as to warm, moisten, and clean the inspired air [26]. Gas exchange of oxygen and carbon dioxide occurs at the alveolar surface (43–102 m^2) [27], which has a considerably larger surface area than the conducting airways (2.5 m^2) [19, 28, 29].

The lung epithelium is made up of a monolayer of cells. The types and structures of the cells are different in the conducting airways and alveolar space [21]. The epithelium of the conducting airways is made of pseudo-stratified columnar epithelia [30] and populated with different cells, such as basal, ciliated, and secretory cells (goblet, mucous, serous, and Clara cells), to maintain the integrity and hemostasis

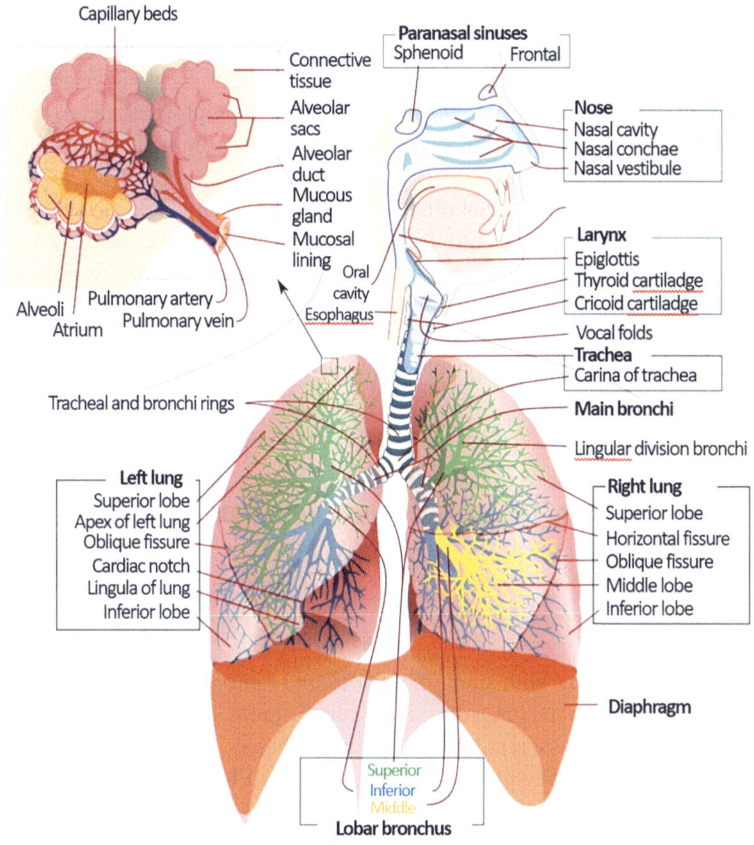

Fig. 2 Lung physiology. (Adapted from Williams [22])

of the respiratory tract [31, 32]. The alveolar epithelium is composed of a single layer of Type I and Type II cells. Type I cells are very large squamous cells (50–100 μm in diameter) with thin cytoplasmic extensions (average cell thickness of 0.26 μm). Type II cells are small, cuboidal cells that possess specialized organelles (0.1–2.4 μm in diameter) called lamellar bodies, which contain surfactants [30]. The diameters and thickness of the cells decrease with a gradual decline into the distal airways [25, 33, 34].

On the apical surfaces of the conducting and respiratory airways is a layer of epithelial lining fluid (ELF), which is populated by a monolayer of surfactants, with a hydrophobic fatty acid tail projected into the airway lumen [24, 25]. The roles of the pulmonary surfactants are to reduce surface tension and facilitate gas exchange [35]. The ELF is slightly hypotonic or isotonic and slightly acidic (pH 6.9), containing a small amount of proteins, such as α1-antitrypsin, immunoglobulins, albumin, and non-immunoglobulin proteins [36]. The thickness of the ELF decreases from 5–10 μm in the conducting airways to 0.05–0.08 μm in the alveoli. The lung interstitium is made of interstitial fluid, connective fibers, and different types of cells (e.g., fibroblasts, monocytes, and lymphocytes) [24, 25, 37].

The lungs have a dual blood supply system: the bronchial and pulmonary circulatory systems. Pulmonary circulation is low pressure and high volume, which facilitates efficient gas exchange. The entire cardiac output from the right heart is perfused into the alveolar capillaries via this circulation. High-pressure, low-volume bronchial circulation is part of the systemic circulation, from which it receives only about 1% of the cardiac output [38, 39].

For drugs to be aerosolized/nebulized, they must be formulated as either a liquid or solid aerosol. In the case of biopharmaceuticals, this requires delicate formulation engineering and manufacturing [40]. The therapeutic agent needs to match the desired therapeutic outcome, taking into consideration the overall host system's biological barriers (deposition and clearance mechanisms) or targets (receptors or pathogens) to guide the selection or development of a suitable inhalational drug delivery system [41].

2.2 *Pulmonary PK Processes: Absorption, Distribution, Metabolism and Elimination (ADME)*

2.2.1 Pulmonary Deposition

Inhaled drugs (including both small molecules and biological macromolecules) are usually administered to the lungs through three different kinds of aerosol inhalation devices: nebulizers, pressurized metered dose inhalers (pMDI), and dry powder inhalers (DPI) [42, 43]. The pulmonary and systemic PK of inhaled drugs is highly dependent on pulmonary deposition patterns [44, 45]. The large variance in drug deposition in the lungs from DPIs is due to many factors, such as inhalation techniques, formulation, density, particle size, aerosol velocity and geometry, and lung morphology [11]. Particles with aerodynamic diameters between 2 and 10 μm are deposited in the tracheobronchial tree, while those with greater diameters are deposited in the extra-thoracic region (the nose, mouth, pharynx, and larynx) [42]. However, particles that are smaller than 1 μm are likely to be exhaled if they do not adhere to the cell walls and/or aggregation does not occur, while particles with a size range of 1–2 μm tend to be distributed into the capillary-rich alveolar region [42, 46]. Furthermore, patient-specific factors and disease statuses have been demonstrated to impact the deposition patterns of drugs, and subsequently, the pulmonary and systemic PK [47]. Lin et al. offer a comprehensive review of factors affecting particle deposition following inhalation [42].

2.2.2 Pulmonary Dissolution

Inhalation can be used for both local or systemic delivery of drugs. For locally-acting drugs, inhaled particles need to dissolve in ELF in order to elicit the pharmacological effect. For systemically-acting drugs, the drugs achieve optimal absorption into the systemic circulation to elicit the pharmacological effect. Before pulmonary absorption can occur, inhaled drug particles deposited in the lungs must dissolve in the ELF. The dissolution process is dependent on the physiochemical properties of the drug/formulation (e.g., its solubility) and the physiology of the lung (e.g., the volume of ELF at the site of deposition) [42]. Furthermore, the ELF where the inhaled particles dissolve varies regionally in thickness, composition, and volume. For example, the mucus gel in the trachea, bronchi, and bronchioles is ~3–23 μm and consists of 95% water, 2–3% mucins, 0.3–0.25% lipids, 0.1–0.5% non-mucin proteins, and other cellular debris. However, the film of surfactant in the alveolar region (~0.07 μm) contains 90% lipids and 10% surfactant proteins [48]. The dissolution process is likely to be the same for both small molecules and biological macromolecules.

2.2.3 Absorption

Following the dissolution of the inhaled drugs in the airway, there are several barriers to their absorption into the systemic circulation. The alveolar surfactant or bronchial mucus layer and the fluid covering the epithelial surface of the lung form further barriers to absorption following drug deposition [11, 49]. The alveolar surfactant may increase the solubility of small, lipophilic drug molecules but induces the aggregation of large molecules and potentially compromises drug dissolution or enhances macrophage digestion. The mucus layer of the conducting airways forms a mesh of hydrophobic and negatively charged sites. Bioavailability can be increased by enhancing mucus adhesion (e.g., chitosan coating on particles) [50]. The most significant barrier to the absorption of inhaled drugs is generally the epithelium of the lung, as the permeability of the epithelium is thought to be much lower than that of the pulmonary endothelium [11].

The mechanism of drug transport across the lung epithelium occurs via passive and active transport mechanisms, including paracellular or transcellular transport, pore formation, and vesicular transport. Transcellular diffusion is the most common route for hydrophobic compounds. For hydrophobic drug molecules with molecular weights of 0.1–1 kDa, the rate of absorption is rapid compared to hydrophilic compounds and is determined by lipophilicity [51].

Hydrophilic compounds diffuse across the epithelium via a paracellular route, where they are diffused through aqueous pores in the intercellular junction [52]. For hydrophilic drug molecules, the rate of absorption is determined by the degree of ionization and molecular weight. Ionized molecules are absorbed slower due to increased interaction with proteins and lipids lining the pores [53]. For hydrophilic compounds with molecular weights of 0.1–1 kDa, the degree of ionization predominantly affects the rate of absorption [21, 54]. Given that biological macromolecules are highly charged and hydrophilic, they often interact significantly with mucus and other pulmonary surfactants. More specifically, the rate of absorption for biological macromolecules seems to be determined by the number of FcRn fragments within an antibody [6].

Biological macromolecules (with a molecular weight range of 1–5 kDa) are generally thought to be absorbed through both the intercellular tight junctions and endocytic vesicles through diffusion-limited processes. However, intercellular tight junctions of the respiratory epithelium have a cut-off of 0.5–0.9 nm, while the underlying endothelium allows the movement of particles with diameters of 6.5–7.5 nm. Although the exact mechanisms of biological absorption in the lungs are not completely understood, an inverse relationship between absorption rate and molecular mass has been established [12].

Absorption of inhaled drugs can also be facilitated or hindered by active processes, such as drug transporters. Two major members of the drug transporter family are present in lung epithelial cells: the ATP-binding cassette (ABC transporters) family and the solute carrier (SLC) family [52]. Within the SLC family, organic cation transporters (OCT) and organic anionic transporters (OAT, OATP) are responsible for the transportation of minimally passively permeable drugs [52], while peptide transporters (PEPT2) are involved in the transportation of peptide drugs [55]. On the other hand, ABC transporters, such as breast cancer resistance protein (BCRP) and P-glycoprotein (P-gp/MDR1), are responsible for the efflux of drug molecules out of cells. Generally, biological macromolecules are too large and are unlikely to be substrates of transporters; hence, the absorption profile of most biological macromolecules is likely to be independent of the transporters [52, 56, 57]. Nevertheless, there are several biological macromolecules (e.g., albumin [58], alpha-1 antitrypsin (AAT) [59] and transferrin [60]) that are found to be actively absorbed through receptor-mediated transcytosis. FcRn-mediated transcytosis was found to be the predominant receptor-mediated transcytosis for biological macromolecules in the upper and central respiratory tract.

Given the heterogeneous nature of the cells along the human respiratory tract, the absorption rates of drugs differ in different areas of the lung. Inhaled particle deposits in the upper airway tend to dissolve faster due to the large volume of fluid in the upper airway. However, absorption in the upper airways is often poor due to the thicker airway wall. On the other hand, peripheral airways have a significantly larger surface area and thinner wall that would allow greater absorption into the systemic circulation but at the same time, drug dissolution is poor due to a smaller volume of ELF. Despite the direct correlation between the bioavailability of the inhaled drug with the deposition depth in the airways, the optimal site for the deposition of inhaled biological macromolecules remains unclear and is likely to be treatment-dependent [6]. If the goal is to treat a systemic disease, then it appears to be important for the drugs to deposit in the alveolar space for optimal absorption [11].

2.2.4 Distribution

The PK of inhaled drugs is dependent on the accumulation and distribution characteristics of the human lungs. The distribution or accumulation of a small molecule in the lung is highly dependent on the degree of protonation of the nitrogen atom (i.e., the degree of ionization) and the lipophilicity [11]. Non-basic amines (pKa < 7.0) have been shown to accumulate to a lesser extent in rabbit lungs in comparison to lipophilic basic amines (pKa > 8.5), where extensive binding to human and rabbit lungs has been observed [61, 62]. Binding of lipophilic basic amines can occur via two mechanisms: saturable and non-saturable. The binding of small molecules to intracellular binding sites, such as mitochondria and lysosomes, usually occurs via carrier-mediated transport systems, representing the saturable mechanism [63]. In contrast, the non-specific binding of small molecules in the extracellular fluid represents the non-saturable mechanism [64]. Our current understanding of the pulmonary distribution or accumulation of inhaled biological macromolecules remains very limited. Given that biological macromolecules have many ionizable groups as well as complex ionization states, the mechanisms governing the distribution and accumulation of biological macromolecules are expected to be more complicated than those described earlier for small drug molecules. Further studies are urgently needed to elucidate the mechanisms governing the pulmonary distribution and accumulation of biological macromolecules following inhalation.

2.2.5 Metabolism

Enzyme-limited clearance and susceptibility to the induction or inhibition of metabolic enzymes is likely, as the intrinsic clearance of drugs in the lungs is relatively low [65]. Although the lungs have lower levels of drug metabolism enzyme classes compared to the liver, many cell types in the lungs, such as Clara cells, alveolar type II cells, and macrophages, are involved in metabolizing xenobiotic compounds [65]. The expression levels and patterns of the CYP enzymes in the lungs are different from those of the liver and intestines [52]. The most common Phase I drug-metabolizing enzymes expressed in the lungs are CYP1A1, CYP1B1, CYP2A6, CYP2B6/7, CYP2E1, CYP2J2, CYP3A4, CYP3A5, epoxide hydrolase (EH), flavine mono-oxygenases (FMO). Phase II metabolizing enzymes include sulfotransferases (SULT), UDP glucuronosyl transferases (UGT), esterases, peptidases, and cyclo-oxygenases [65–67].

Small natural peptides (<3 kDa) are more affected by degradation in the respiratory tract than larger (5–500 kDa) proteins [68]. Generally, proteins with a molecular weight range of 6–50 kDa are not affected by peptidases and tend to have high bioavailability [6]. The effects of inhaled proteins and peptides are also influenced by metalloproteinase (MMP)-2 in the hypophase and plasma membrane-bound carboxypeptidase-M of alveolar epithelial type 1 cells. Alveolar macrophages perform functions in the nonspecific immune system by removing undissolved particles through uptake, while peroxidases and proteases secreted by the immune cells reduce protein absorption. As macrophages preferentially take up proteins as protein aggregates or in denatured form, more than 95% of proteins are absorbed intact from the lungs [12].

2.2.6 Elimination

Three distinct types of elimination or clearance mechanisms exist in the lungs. The first clearance mechanism is mucociliary clearance, which occurs mostly in the conducting airways and is relevant only for undissolved particles. Here, ciliated epithelium beats to facilitate the movement of the particles trapped in the mucus out of the lung and into the trachea; eventually, they are swallowed into the gastrointestinal tract (GIT). Mucociliary clearance typically occurs when undissolved drug particles or protein aggregates remain in the mucus [55]. The second clearance mechanism is macrophage uptake, which is the main clearance mechanism in the alveoli for undissolved particles. Inhaled drugs are engulfed by alveolar macrophages in the deep lungs via pinocytosis in a size-dependent manner [11, 52]. Particles with diameter less than 1 μm can readily avoid phagocytosis, while those with diameters between 1 and 2 μm are readily phagocytosed by alveolar macrophages. Following uptake, the macrophage undergoes internal enzymatic degradation or is drained into the lymphatic system. The third clearance mechanism is absorption into the systemic circulation. After deposition in the alveolar region, drug particles may dissolve in the ELF and become available for systemic absorption [11, 69]. All three clearance mechanisms occur simultaneously, and the type of clearance mechanism a drug is subjected to depends on the deposition site. Finally, the clearance pathway competes with the epithelial transport in the airways [6]. Biological macromolecules with slow absorption rates are retained in the airways longer which provides sufficient time for drugs to be cleared by one of three clearance mechanisms. On the other hand, systemic absorption of small peptides and proteins, e.g., insulin [70], is less affected by the pulmonary clearance mechanism as they are rapidly absorbed into the systemic circulation.

2.2.7 Key Factors That Determine the PK of Inhaled Biological Macromolecules

In summary, biological macromolecules with a molecular weight below 40 kDa are rapidly absorbed into the systemic circulation but are also subject to substantial metabolism in the airways [6]. On the contrary, biological macromolecules with a molecular weight greater than 40 kDa are not metabolized to the same extent as the smaller biological macromolecules but are poorly absorbed in the systemic circulation.

2.3 Animal-Based Inhalation Model for Clinical Development of Inhaled Biological Macromolecules

Preclinical inhalation studies, which typically investigate the PK and PD of inhaled drugs, are required by regulatory authorities prior to initiating clinical studies. Animal-based inhalation studies are often carried out with rodents and rabbits in restraint tubes and with dogs and primates using face masks or helmets to mimic human behavior [71].

Animal-based preclinical studies have several limitations. The first is associated with animal models, as lung physiology and anatomy differ across species, likely causing different particle deposition patterns. Rodents' lung anatomy differs vastly from that of humans, while monkeys, baboons, and guinea pigs have lung anatomy that closely mimics it. Interspecies differences are summarized and discussed in Wylie et al. [72]. The second limitation is associated with inhalation devices, as breath-actuated devices, such as DPIs, cannot be used in preclinical animal models. To overcome this issue, animals are exposed by means of specialized equipment used to aerosolize the solution or powder [1].

Different types of devices are utilized across different animal models and stages of drug development. In early-stage drug development with rodents, microsprayers and/or insufflators are commonly used [73]. Microsprayers are syringe-like handles that can be filled with a drug solution or suspension, while insufflators are designed to deliver dry powders. Both devices connect to a needle-like component that is inserted into an animal's trachea to deliver the drug. These two delivery techniques have several drawbacks. One is that they disperse powder and aerosolize particles differently from clinical nebulizers and dry powder inhalers [72]. Many studies have demonstrated that both devices result in uneven drug distribution in the lungs [74]. Inserting these devices into the lung can also induce inflation, which likely impacts the pulmonary PK processes. Later stages of drug discovery and development tend to employ true inhalation systems, such as whole-body chambers [75]. The utility of these devices varies depending on the species [76]. Readers are referred to Wylie et el. for an excellent overview of the types of devices available at different stages of drug development [72]. It is important to note that late-stage devices also have some drawbacks, e.g., the test agent will deposit on the fur and since it is hard to control drug exposure precisely, PK/PD analysis needs to be completed using sophisticated computational software [75].

In any drug development program, the transition from the preclinical to the clinical stage is challenging. This is especially true for inhalation programs due to their complex kinetic processes. The selection of an inhalation device requires complex considerations and will significantly impact the interpretation of the results. Modeling and simulation techniques can be utilized to bridge the gap between the preclinical and clinical stages, thus facilitating a smooth transition.

2.4 Modeling and Simulation Tools

Model-informed drug development (MIDD), also known as modeling and simulation, is unarguably the cornerstone of pharmacological research in the twenty-first century [77]. It refers to the strategic creation and integration of mathematical models with thorough plans of execution (e.g., key questions, assumptions, a modeling approach, and documentation) to facilitate decision-making in pharmaceutical research [78–80]. MIDD applications include novel target identification, formulation design, nonclinical and clinical development, biopharmaceutical research, trial design, and cost-effectiveness evaluations. MIDD is also increasingly used to evaluate causal links among drug physiochemical properties, disease/pathogen biology, and patient physiology. The method offers an integrated approach for designing effective trials and transforming data to knowledge, thus, narrowing knowledge gaps and maximizing the therapeutic potential of inhaled drug candidates. Modeling and simulation are extensively employed for predicting the PK and PD of inhaled therapeutics [77, 81]. Several reviews have examined deposition models, empirical PK models, and physiologically-based pharmacokinetic (PBPK) models for inhaled therapeutics [18, 82, 83]. Unfortunately, to the best of our knowledge, no publications have used these models for inhaled biological macromolecules; therefore, detailed assessments of their utility or performance for these medications are unable to be provided.

2.4.1 Deposition Models

The deposition pattern of inhaled drugs in respiratory airways is key in determining the delivered dose and the kinetic processes. Particles that deposit on the walls of the throat and mouth will never reach the lung, and aerodynamically small particles (<5 μm) will likely be exhaled and unable to reach the target site [42]. For inhalation therapy to be successful, therefore, the aerosolized particles must not deposit in the throat or mouth and must have an aerodynamic diameter suitable for effective deposition in lung airways [84]. Mathematical modeling of a deposition pattern can help elucidate the complex kinetic processes following inhalation. Readers are referred to a recent critical overview of state-of-the-art deposition modeling [85].

Empirical Deposition Correlations Based on In Vivo Data

Empirical correlations, the simplest mathematical models, are derived from in vivo data and typically used to predict total drug deposition or the regional drug deposition fraction within a defined region (e.g., the upper airways). As they are based on actual in vivo data, these models tend to yield results consistent with observations. However, given the models' empirical nature, their results cannot be extrapolated beyond the data range upon which the correlations were based. The advantages and limitations associated with the empirical correlation methods have been discussed in detail elsewhere [86].

Generational Deposition Models

Empirical correlations are only feasible when in vivo data are available. When they are not, more complex mechanistic deposition models are employed to predict regional deposition to the level of individual airway generation. Readers are referred to a recent review by Fröhlich and Salar-Behzadi for details [12]. In brief, generational deposition models use a simplified representation of the human lungs and mathematical expressions to predict deposition probabilities based on various deposition mechanisms. Many determinants that influence deposition, including airway geometry, breathing patterns, inhalation maneuvers, and particle properties, can be incorporated into these models to predict their likely impact on overall or regional deposition patterns. These models yield a list of drug mass deposition fractions over each lung generation. Deposition models can be combined with PK models (see the section below on PBPK models) to predict regional pulmonary exposure and systemic exposure over time [83].

2.4.2 PK Models

Physiologically-Based PK Models

PBPK models comprise a quantitative approach that incorporates mechanistic information – such as human anatomy and physiology (including intrinsic and extrinsic patient factors) – and a drug's physiochemical properties (e.g., solubility, pKa values, partition coefficients, and dissolution rates) to predict and define PK (e.g., dissolution, absorption, distribution, metabolism, and elimination) [18, 44, 83, 87, 88]. Unlike the data-driven PK/PD approach, PBPK models are based on a series of differential equations that are parameterized with a set of predefined physiologically driven parameters [18]. Readers are referred to Zhuang and Lu for a critical overview of PBPK models [89]. Generic PBPK inhalation models have been implemented into the commercial software programs GastroPlus™ and PulmoSim™ [18, 85]. PBPK models have been widely used to describe pulmonary/systemic PK for small molecules only [18, 85]. To date, there is very limited evidence on the utility of PBPK models for inhaled biological macromolecules.

The GastroPlus™ Additional Dosage Routes Module

The Pulmonary Compartmental Absorption and Transit (PCAT) model included in GastroPlus™ divides the lung into four compartments: the extra-thoracic (naso- and oro-pharynx and larynx), thoracic (trachea and bronchi), bronchiolar (bronchioles and terminal bronchioles) and alveolar-interstitial (respiratory bronchioles, alveolar ducts and sacs, and interstitial connective tissue) (Fig. 3) [18, 85, 90]. The amount of drug deposited in each compartment can be defined manually (via experimental data) or calculated by a built-in particle deposition model developed by the International Commission on Radiological Protection (ICRP 66 model) [18, 85].

The particle kinetics depend on the regional deposition site. Particles that deposit in the extra-thoracic compartment can be either swallowed and absorbed through the GIT – circumstances handled via an advanced compartmental absorption and transit absorption model – or absorbed directly into systemic circulation via a passive process [18, 85]. The dissolution processes of particles that deposit in the other

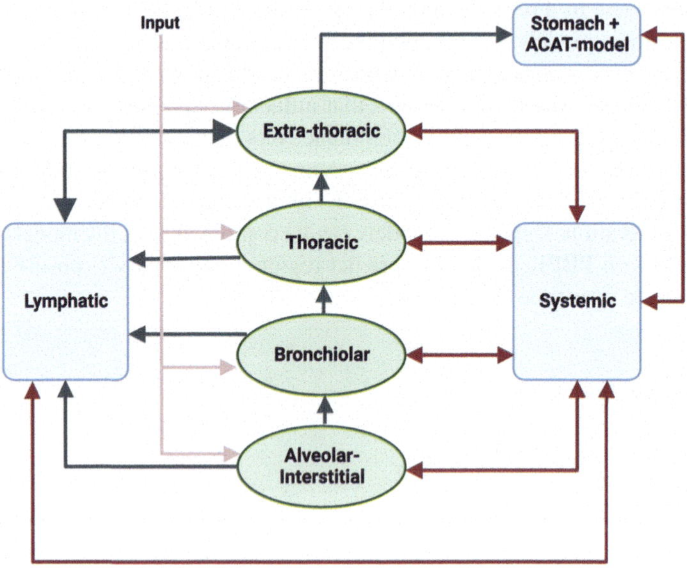

Fig. 3 Schematic of the Pulmonary Compartmental Absorption and Transit (PCAT) model. (Adapted from Borghardt et al. [18])

three pulmonary compartments are handled via different methods, such as the Noyes–Whitney function [18, 85]. Pulmonary absorption from each compartment is assumed to be passive, and rates depend on the pulmonary system's anatomical and physiological parameters (e.g., mucociliary clearance) and drug-dependent input parameters. It is important to note that because the physiological parameters needed for PBPK modeling are difficult to obtain in humans, the values reported in the literature vary widely, making PBPK models difficult to develop. Before utilizing these models to predict or translate, it is thus important to demonstrate in silico-in vivo correlation and validate using clinical data on inhaled drugs [18, 85].

PulmoSim™

The PulmoSim™ PBPK model (Fig. 4) differentiates two routes of an inhaled drug [18, 91]. One fraction of the drug is swallowed and absorbed by the GIT, and the rest is deposited in the lungs, where particles first undergo dissolution. Both dissolved and undissolved drug can be moved via mucociliary clearance to the gut compartment but dissolved drug that escapes this clearance is absorbed by lung tissue before being absorbed into the systemic circulation. Furthermore, the unbound fraction of a drug in the lungs appears to be a key determinant of the drug's absorption from the lungs to the plasma. The PulmoSim™ PBPK model has the same limitations as the PCAT model described above.

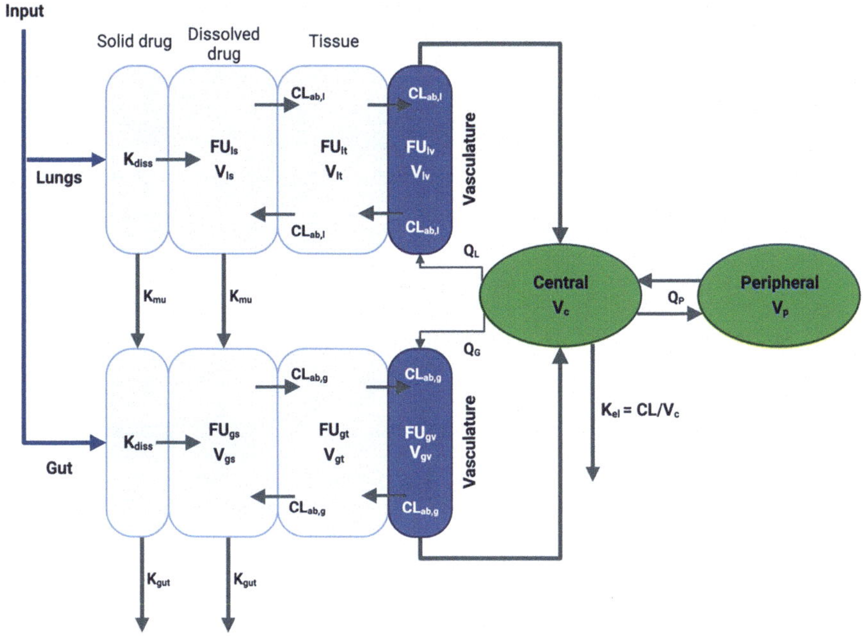

Fig. 4 Schematic of the PulmoSim™. l = lung; g = gut; s = solution; t = tissue; v = vasculature; k_{diss} = dissolution rate constant; k_{mu} = mucociliary rate constant; k_{gut} = elimination rate constant from the gut; k_{el} = elimination rate constant from the central compartment; FU fraction unbound; CL_{ab} = absorptive clearance across tissue; V = physiological distribution volumes; QG = flow rate to the gut vasculature; QL = flow rate to the pulmonary vasculature; QP = intercompartmental clearance; light purple boxes = pulmonary compartments; light purple boxes = gut compartments; dark purple boxes = vasculature compartments in the specific organs; green boxes = systemic disposition compartments. (Adapted from Borghardt et al. [18])

Data-Driven PK/PD Approaches

Unlike the mechanistic PBPK modeling approach, data-driven approaches rely largely on available preclinical or clinical PK data, selecting model structures based on pharmacostatistical methods and biological plausibility [92]. Data-driven models are often less complex than PBPK models, and their parameters often carry less biologically meaningful information. One major advantage of data-driven models is that they can identify important predictors to explain PK characteristics.

The literature includes many different types of pulmonary absorption models [18], which can be generally categorized into two main absorption model types, the choice of which is largely dependent on the PK data. When no PK data from intravenous (IV) or other administration routes are available for model development, Model I (Fig. 5) is preferred. When IV data are available, complexity increases, making Models II, IIIa, and IIIb preferable.

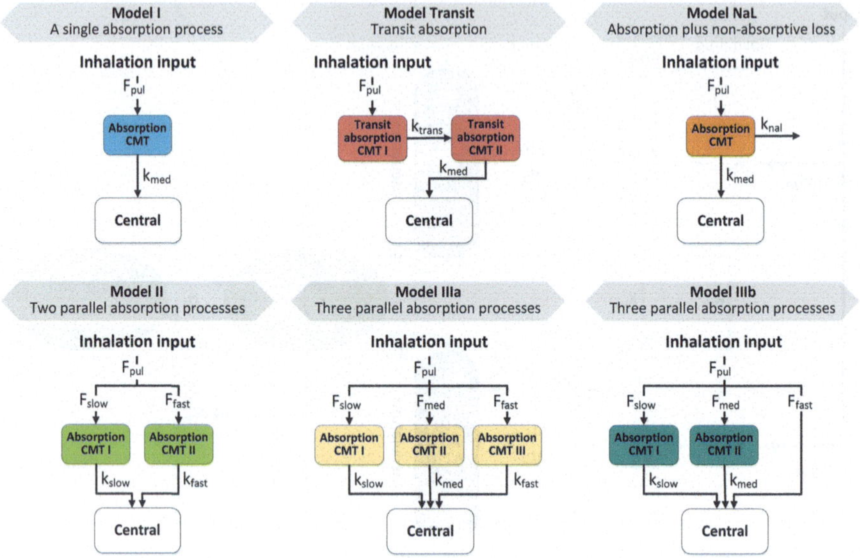

Fig. 5 Structural models for pulmonary absorption. (Adapted from Himstedt et al. [87]). CMT = compartment; F_{Pul} = pulmonary bioavailability or designated lung dose; $F_{slow/med/fast}$ = fraction of the lung dose slowly/intermediately/fastly absorbed; $k_{slow/med/fast}$ = slow/intermediate/fast absorption rate constants; k_{trans} = transit rate constant; k_{nal} = non-absorptive loss rate constant

Another major advantage of data-driven empirical PK/PD modeling is that it can be used to establish a quantitative link between exposure (i.e., PK) and the efficacy of an inhaled drug [93]. Recently, a mechanism-based PK/PD model (MBM) for inhaled colistin was developed based on PK/PD data of infected neutropenic mice. It was subsequently combined with a previously published PK model of nebulized colistin methanesulfonate to predict the treatment outcomes of different clinically prescribed dosage regimens in patients. The MBM had previously successfully estimated the PK/PD targets observed for inhaled colistin in neutropenic mice. In line with previous clinical recommendations, deterministic simulation employing human PK data predicted that an inhalation dosage regimen of 60 mg colistin base activity every 12 hours is more likely to achieve a ≥2-log10 CFU/lung bacterial reduction in critically ill patients with lung infections caused by multidrug-resistant *Pseudomonas aeruginosa* [93]. This recently developed MBM is a valuable and clinically useful tool for optimizing inhalational dosage regimens.

3 Summary

The development of inhaled biological macromolecules has expanded significantly over the last decade, making pulmonary delivery a promising alternative to systemic delivery of biological macromolecules. However, pulmonary deposition and ADME

Table 2 Challenges of PBPK modeling

No.	Challenges for PBPK modeling
1	Shortage of individuals and independent reviewers with M&S experience and expertise.
2	The gap between available models and physiochemical and physiological processes.
3	Limitations in experiments to estimate the desired parameters.
4	Limited available in vivo study data, and difficulty in sampling in studies. This results in a lack of confidence in model extrapolation.
5	Lack of information on intra- and inter-subject variability. Difficult to incorporate physiological differences between normal and special populations into the PBPK models.
6	Knowledge gap in the physiochemical properties of the formulation, PK/PD behaviors, and their representation by in silico models.
7	Models coded in one M&S platform cannot be easily translated to a different M&S platform.

processes are highly complex, and their interplay determines the pulmonary and systemic PK/PD of inhaled biological macromolecules. Although great progress has been made in understanding pulmonary deposition and ADME processes, a significant knowledge gap remains, making modeling of pulmonary kinetic processes challenging (Table 2). In addition, the literature varies widely regarding the parameters needed for PBPK modeling. Well-designed in vitro experiments, clinical studies, and more sophisticated modeling approaches are urgently needed to elucidate the highly complex pulmonary deposition and ADME processes for the clinical development of inhaled medications.

Conflict of Interest The authors declare that they have no known financial or interpersonal conflicts that would have appeared to have an impact on the research presented in this study. Dr Yu-Wei Lin is currently an employee of Certara, USA. Ms. Audrey Huili Lim is currently employed by the Institute for Clinical Research, National Institutes of Health, Shah Alam, Malaysia.

References

1. Movia D, Prina-Mello A. Preclinical development of orally inhaled drugs (OIDs)—are animal models predictive or shall we move towards in vitro non-animal models? Animals. 2020;10:1259.
2. WHO. Chronic obstructive pulmonary disease (COPD). Available online: https://www.who.int/news-room/fact-sheets/detail/chronic-obstructive-pulmonary-disease-. Accessed on 23 Mar 2023.
3. WHO. Asthma. Available online: https://www.who.int/news-room/q-a-detail/asthma. Accessed on 23 Mar 2023.
4. WHO. The top 10 causes of death 2020. Available from: https://www.who.int/news-room/fact-sheets/detail/the-top-10-causes-of-death.
5. GBI Research. Global respiratory drugs market to 2023—a changing therapeutic landscape as key patents expire and biologics, targeted therapies and CFTR modulators for asthma and cystic fibrosis treatment emerge as market growth drivers. Available online: http://gbiresearch.

com/report-store/market-reports/therapy-analysis/global-respiratory-drugs-market-to2023-a-changing-therapeutic-landscape-as-key-patents-expire-and-biologics-targeted-therapi. Accessed on 26 Mar 2022.

6. Liang W, Pan HW, Vllasaliu D, Lam JKW. Pulmonary delivery of biological drugs. Pharmaceutics. 2020;12(11):1025.
7. Jozala AF, Geraldes DC, Tundisi LL, Feitosa VA, Breyer CA, Cardoso SL, et al. Biopharmaceuticals from microorganisms: from production to purification. Braz J Microbiol. 2016;47(Suppl 1):51–63.
8. Pham JV, Yilma MA, Feliz A, Majid MT, Maffetone N, Walker JR, et al. A review of the microbial production of bioactive natural products and biologics. Front Microbiol. 2019;10:1404.
9. Adami G, Saag KG, Chapurlat RD, Guañabens N, Haugeberg G, Lems WF, et al. Balancing benefits and risks in the era of biologics. Ther Adv Musculoskelet Dis. 2019;11:1759720x19883973.
10. Mishra B, Singh J. Novel drug delivery systems and significance in respiratory diseases. In: Targeting chronic inflammatory lung diseases using advanced drug delivery systems. London: Academic Press; 2020. p. 57–95.
11. Labiris NR, Dolovich MB. Pulmonary drug delivery. Part I: physiological factors affecting therapeutic effectiveness of aerosolized medications. Br J Clin Pharmacol. 2003;56(6):588–99.
12. Fröhlich E, Salar-Behzadi S. Oral inhalation for delivery of proteins and peptides to the lungs. Eur J Pharm Biopharm. 2021;163:198–211.
13. Erdő F, Bors LA, Farkas D, Bajza Á, Gizurarson S. Evaluation of intranasal delivery route of drug administration for brain targeting. Brain Res Bull. 2018;143:155–70.
14. Katzung BG, Masters SB, Trevor AJ. Basic & clinical pharmacology. New York/London: McGraw-Hill; 2004.
15. Orme M, Sjöqvist F, Birkett D, Brøsen K, Cascorbi I, Gustafsson LL, et al. Clinical pharmacology in research, teaching and health care. Basic Clin Pharmacol Toxicol. 2010;107(1):531–59.
16. Rodríguez-Gascón A, Solinís MÁ, Isla A. The role of PK/PD analysis in the development and evaluation of antimicrobials. Pharmaceutics. 2021;13(6):833.
17. Tuntland T, Ethell B, Kosaka T, Blasco F, Zang RX, Jain M, et al. Implementation of pharmacokinetic and pharmacodynamic strategies in early research phases of drug discovery and development at Novartis Institute of Biomedical Research. Front Pharmacol. 2014;5:174.
18. Borghardt JM, Weber B, Staab A, Kloft C. Pharmacometric models for characterizing the pharmacokinetics of orally inhaled drugs. AAPS J. 2015;17(4):853–70.
19. Powers KA, Dhamoon AS. Physiology, pulmonary ventilation and perfusion. In: StatPearls [Internet]. Treasure Island: StatPearls Publishing; 2022.
20. Haddad M, Sharma S. Physiology, lung. In: StatPearls [Internet]. Treasure Island: StatPearls Publishing; 2022.
21. Liu X, Jin L, Upham JW, Roberts MS. The development of models for the evaluation of pulmonary drug disposition. Expert Opin Drug Metab Toxicol. 2013;9(4):487–505.
22. Williams PL. Gray's anatomy: the anatomical basis of medicine and surgery. 38th ed. London: Churchill Livingstone; 1996. 2092 p.
23. Van de Moortele T, Wendt CH, Coletti F. Morphological and functional properties of the conducting human airways investigated by in vivo computed tomography and in vitro MRI. J Appl Physiol (1985). 2018;124(2):400–13.
24. Patton JS. Mechanisms of macromolecule absorption by the lungs. Adv Drug Deliv Rev. 1996;19(1):3–36.
25. Patton JS, Byron PR. Inhaling medicines: delivering drugs to the body through the lungs. Nat Rev Drug Discov. 2007;6(1):67–74.
26. Bur M, Lehr C-M. Pulmonary cell culture models to study the safety and efficacy of innovative aerosol medicines. Expert Opin Drug Deliv. 2008;5(6):641–52.
27. Weibel ER. Morphological basis of alveolar-capillary gas exchange. Physiol Rev. 1973;53(2):419–95.
28. Hsia CC, Hyde DM, Weibel ER. Lung structure and the intrinsic challenges of gas exchange. Compr Physiol. 2016;6(2):827–95.

29. Mercer RR, Russell ML, Roggli VL, Crapo JD. Cell number and distribution in human and rat airways. Am J Respir Cell Mol Biol. 1994;10(6):613–24.
30. Mathias NR, Yamashita F, Lee VHL. Respiratory epithelial cell culture models for evaluation of ion and drug transport. Adv Drug Deliv Rev. 1996;22(1):215–49.
31. Gartner LP. Respiratory system textbook of histology. 2nd ed. Elsevier Inc.; 2021. p. 355–78.
32. Kia'i N, Bajaj T. Histology, respiratory epithelium. In: StatPearls [Internet]. Treasure Island: StatPearls Publishing; 2022.
33. Ward HE, Nicholas TE. Alveolar type I and type II cells. Aust NZ J Med. 1984;14(5 Suppl 3):731–4.
34. Ruaro B, Salton F, Braga L, Wade B, Confalonieri P, Volpe MC, et al. The history and mystery of alveolar epithelial type II cells: focus on their physiologic and pathologic role in lung. Int J Mol Sci. 2021;22(5):2566.
35. King RJ. Pulmonary surfactant. J Appl Physiol Respir Environ Exerc Physiol. 1982;53(1):1–8.
36. Rennard SI, Basset G, Lecossier D, O'Donnell KM, Pinkston P, Martin PG, et al. Estimation of volume of epithelial lining fluid recovered by lavage using urea as marker of dilution. J Appl Physiol (1985). 1986;60(2):532–8.
37. Brandsma J, Goss VM, Yang X, Bakke PS, Caruso M, Chanez P, et al. Lipid phenotyping of lung epithelial lining fluid in healthy human volunteers. Metabolomics. 2018;14(10):123.
38. Suresh K, Shimoda LA. Lung circulation. Compr Physiol. 2016;6(2):897–943.
39. Jain V, Bordes SJ, Bhardwaj A. Physiology, pulmonary circulatory system. In: StatPearls [Internet]. Treasure Island: StatPearls Publishing; 2022.
40. Scherließ R, Bock S, Bungert N, Neustock A, Valentin L. Particle engineering in dry powders for inhalation. Eur J Pharm Sci. 2022;172:106158.
41. Hickey AJ. Emerging trends in inhaled drug delivery. Adv Drug Deliv Rev. 2020;157:63–70.
42. Lin YW, Wong J, Qu L, Chan HK, Zhou QT. Powder production and particle engineering for dry powder inhaler formulations. Curr Pharm Des. 2015;21(27):3902–16.
43. Hickey AJ. Pharmaceutical inhalation aerosol technology. Boca Raton: CRC Press; 2003.
44. Hartung N, Borghardt JM. A mechanistic framework for a priori pharmacokinetic predictions of orally inhaled drugs. PLoS Comput Biol. 2020;16(12):e1008466.
45. Hochhaus G, Horhota S, Hendeles L, Suarez S, Rebello J. Pharmacokinetics of orally inhaled drug products. AAPS J. 2015;17(3):769–75.
46. Chow AH, Tong HH, Chattopadhyay P, Shekunov BY. Particle engineering for pulmonary drug delivery. Pharm Res. 2007;24(3):411–37.
47. Chaurasiya B, Zhao YY. Dry powder for pulmonary delivery: a comprehensive review. Pharmaceutics. 2020;13(1):31.
48. Eedara BB, Bastola R, Das SC. Dissolution and absorption of inhaled drug particles in the lungs. Pharmaceutics. 2022;14(12):2667.
49. Fahy JV, Dickey BF. Airway mucus function and dysfunction. N Engl J Med. 2010;363(23):2233–47.
50. Shin GH, Kim JT. Comparative study of chitosan and oligochitosan coatings on mucoadhesion of curcumin nanosuspensions. Pharmaceutics. 2021;13(12):2154.
51. Alqahtani MS, Kazi M, Alsenaidy MA, Ahmad MZ. Advances in oral drug delivery. Front Pharmacol. 2021;12:618411.
52. Olsson B, Bondesson E, Borgström L, Edsbäcker S, Eirefelt S, Ekelund K, et al. Pulmonary drug metabolism, clearance, and absorption. In: Smyth HDC, Hickey AJ, editors. Controlled pulmonary drug delivery. New York: Springer; 2011. p. 21–50.
53. Patton JS, Fishburn CS, Weers JG. The lungs as a portal of entry for systemic drug delivery. Proc Am Thorac Soc. 2004;1(4):338–44.
54. Ozsoy Y, Gungor S, Cevher E. Nasal delivery of high molecular weight drugs. Molecules. 2009;14(9):3754–79.
55. Groneberg DA, Fischer A, Chung KF, Daniel H. Molecular mechanisms of pulmonary peptidomimetic drug and peptide transport. Am J Respir Cell Mol Biol. 2004;30(3):251–60.

56. Lin L, Yee SW, Kim RB, Giacomini KM. SLC transporters as therapeutic targets: emerging opportunities. Nat Rev Drug Discov. 2015;14(8):543–60.
57. Nigam SK. What do drug transporters really do? Nat Rev Drug Discov. 2015;14(1):29–44.
58. John TA, Vogel SM, Minshall RD, Ridge K, Tiruppathi C, Malik AB. Evidence for the role of alveolar epithelial gp60 in active transalveolar albumin transport in the rat lung. J Physiol. 2001;533(Pt 2):547–59.
59. Lockett AD, Brown MB, Santos-Falcon N, Rush NI, Oueini H, Oberle AJ, et al. Active trafficking of alpha 1 antitrypsin across the lung endothelium. PLoS One. 2014;9(4):e93979.
60. Kim KJ, Malik AB. Protein transport across the lung epithelial barrier. Am J Physiol Lung Cell Mol Physiol. 2003;284(2):L247–59.
61. Boer F. Drug handling by the lungs. Br J Anaesth. 2003;91(1):50–60.
62. Roerig DL, Kotrly KJ, Dawson CA, Ahlf SB, Gualtieri JF, Kampine JP. First-pass uptake of verapamil, diazepam, and thiopental in the human lung. Anesth Analg. 1989;69:461–6.
63. Hallifax D, Houston JB. Saturable uptake of lipophilic amine drugs into isolated hepatocytes: mechanisms and consequences for quantitative clearance prediction. Drug Metab Dispos. 2007;35(8):1325–32.
64. Frutiger A, Tanno A, Hwu S, Tiefenauer RF, Vörös J, Nakatsuka N. Nonspecific binding—fundamental concepts and consequences for biosensing applications. Chem Rev. 2021;121(13):8095–160.
65. Ding X, Kaminsky LS. Human extrahepatic cytochromes P450: function in xenobiotic metabolism and tissue-selective chemical toxicity in the respiratory and gastrointestinal tracts. Annu Rev Pharmacol Toxicol. 2003;43:149–73.
66. Zhang JY, Wang Y, Prakash C. Xenobiotic-metabolizing enzymes in human lung. Curr Drug Metab. 2006;7(8):939–48.
67. Somers GI, Lindsay N, Lowdon BM, Jones AE, Freathy C, Ho S, et al. A comparison of the expression and metabolizing activities of phase I and II enzymes in freshly isolated human lung parenchymal cells and cryopreserved human hepatocytes. Drug Metab Dispos. 2007;35(10):1797–805.
68. Bruno BJ, Miller GD, Lim CS. Basics and recent advances in peptide and protein drug delivery. Ther Deliv. 2013;4(11):1443–67.
69. Guo Y, Bera H, Shi C, Zhang L, Cun D, Yang M. Pharmaceutical strategies to extend pulmonary exposure of inhaled medicines. Acta Pharm Sin B. 2021;11(8):2565–84.
70. Lombry C, Edwards DA, Préat V, Vanbever R. Alveolar macrophages are a primary barrier to pulmonary absorption of macromolecules. Am J Physiol Lung Cell Mol Physiol. 2004;286(5):L1002–8.
71. Pauluhn J, Mohr U. Inhalation studies in laboratory animals--current concepts and alternatives. Toxicol Pathol. 2000;28(5):734–53.
72. Wylie JL, House A, Mauser PJ, Sellers S, Terebetski J, Wang Z, et al. Inhaled formulation and device selection: bridging the gap between preclinical species and first-in-human studies. Ther Deliv. 2018;9(5):387–404.
73. Wong BA. Inhalation exposure systems: design, methods and operation. Toxicol Pathol. 2007;35(1):3–14.
74. Guillon A, Montharu J, Vecellio L, Schubnel V, Roseau G, Guillemain J, et al. Pulmonary delivery of dry powders to rats: tolerability limits of an intra-tracheal administration model. Int J Pharm. 2012;434(1–2):481–7.
75. Phalen RF, Mannix RC, Drew RT. Inhalation exposure methodology. Environ Health Perspect. 1984;56:23–34.
76. Hoppentocht M, Hoste C, Hagedoorn P, Frijlink H, de Boer A. In vitro evaluation of the DP-4M PennCentury (TM) insufflator. Eur J Pharm Biopharm. 2014;88:153–9.
77. Madabushi R, Seo P, Zhao L, Tegenge M, Zhu H. Review: role of model-informed drug development approaches in the lifecycle of drug development and regulatory decision-making. Pharm Res. 2022;39(8):1669–80.
78. Lesko LJ. Perspective on model-informed drug development. CPT Pharmacometrics Syst Pharmacol. 2021;10(10):1127–9.

79. Marshall S, Madabushi R, Manolis E, Krudys K, Staab A, Dykstra K, et al. Model-informed drug discovery and development: current industry good practice and regulatory expectations and future perspectives. CPT Pharmacometrics Syst Pharmacol. 2019;8(2):87–96.
80. Marshall SF, Burghaus R, Cosson V, Cheung SY, Chenel M, DellaPasqua O, et al. Good practices in model-informed drug discovery and development: practice, application, and documentation. CPT Pharmacometrics Syst Pharmacol. 2016;5(3):93–122.
81. Mitra A, Wang Y. Applications of model informed drug development (MIDD) in drug development lifecycle and regulatory review. Pharm Res. 2022;39(8):1663–7.
82. Rostami AA, Campbell JL, Pithawalla YB, Pourhashem H, Muhammad-Kah RS, Sarkar MA, et al. A comprehensive physiologically based pharmacokinetic (PBPK) model for nicotine in humans from using nicotine-containing products with different routes of exposure. Sci Rep. 2022;12(1):1091.
83. Himstedt A, Bäckman P, Borghardt J. Physiologically-based pharmacokinetic modeling after drug inhalation. In: Inhaled medicines optimizing development through integration of in silico. London: Academic Press; 2021. p. 319–58.
84. Darquenne C. Aerosol deposition in health and disease. J Aerosol Med Pulm Drug Deliv. 2012;25(3):140–7.
85. Zou H, Banerjee P, Leung SSY, Yan X. Application of pharmacokinetic-pharmacodynamic modeling in drug delivery: development and challenges. Front Pharmacol. 2020;11:997.
86. Weibelzahl S, Weber G. Advantages, opportunities, and limits of empirical evaluations: evaluating adaptive systems. Künstl Intell. 2002;3:17–20.
87. Himstedt A, Borghardt JM, Wicha SG. Inferring pulmonary exposure based on clinical PK data: accuracy and precision of model-based deconvolution methods. J Pharmacokinet Pharmacodyn. 2022;49(2):135–49.
88. Kandala B, Hochhaus G. Pharmacometrics in pulmonary diseases. In: Schmidt S, Derendorf H, editors. Applied pharmacometrics. 1st ed. New York: Springer; 2014. p. 349–82.
89. Zhuang X, Lu C. PBPK modeling and simulation in drug research and development. Acta Pharm Sin B. 2016;6(5):430–40.
90. Shi C, Ignjatović J, Liu T, Han M, Cun D, Đuriš J, et al. In vitro – in vivo – in silico approach in the development of inhaled drug products: nanocrystal-based formulations with budesonide as a model drug. Asian J Pharm Sci. 2021;16(3):350–62.
91. Collingwood SP, Coe D, Pryde D, Lock R. Respiratory drug discovery, current developments and future challenges: highlights from the Society of Medicines Research Symposium, held on June 14 th, 2012–Horsham, UK. Drugs Future. 2012;37(8):619–25.
92. Tsamandouras N, Rostami-Hodjegan A, Aarons L. Combining the "bottom up"and "top down" approaches in pharmacokinetic modelling: fitting PBPK models to observed clinical data. Br J Clin Pharmacol. 2015;79(1):48–55.
93. Lin YW, Zhou QT, Han ML, Onufrak NJ, Chen K, Wang J, et al. Mechanism-based pharmacokinetic/pharmacodynamic modeling of aerosolized colistin in a mouse lung infection model. Antimicrob Agents Chemother. 2018;62(3):e01965-17.

Index

J. K. W. Lam, P. C. L. Kwok (eds.), *Respiratory Delivery of Biologics, Nucleic Acids, and Vaccines*, AAPS Introductions in the Pharmaceutical Sciences 8,
https://doi.org/10.1007/978-3-031-47567-2

SPRINGER NATURE

GPSR Compliance

The European Union's (EU) General Product Safety Regulation (GPSR) is a set of rules that requires consumer products to be safe and our obligations to ensure this.

If you have any concerns about our products, you can contact us on ProductSafety@springernature.com

In case Publisher is established outside the EU, the EU authorized representative is:

Springer Nature Customer Service Center GmbH
Europaplatz 3
69115 Heidelberg, Germany

The manufacturer's authorised representative in the EU is Springer Nature Customer Service Centre GmbH, Europaplatz 3, 69115 Heidelberg, Germany. If you have any concerns regarding our products, please contact ProductSafety@springernature.com

Printed and bound by CPI Group (UK) Ltd, Croydon, CR0 4YY
01/10/2026
02225899-0003